THE *DICTATORSHIP* OF THE PROLETARIAT

- 2 -

THE
DICTATORSHIP
OF THE
PROLETARIAT
- 2 -

Gerald McIsaac

ISBN: 978-1-963068-59-7 (sc)
ISBN: 978-1-963068-61-0 (hc)
ISBN: 978-1-963068-60-3 (e)

Library of Congress Control Number: 2024919749

INTRODUCTION TO THE
SECOND EDITION

I decided to write a Second Edition to this book, partly because I am not at all satisfied with the First Edition, and partly because the revolutionary situation is developing very quickly. Recent events have to be taken into consideration.

The sad fact is that I am convinced our civilization has passed its peak, and is now in decline. Our civilization is in danger of collapse. This decline must be reversed.

No doubt, there are those who think that this is an exaggeration, that our civilization could never collapse. To such people, may I suggest that you bear in mind that this was precisely what was said of the Roman Empire!

In fact, throughout history, numerous civilizations have come into existence, risen to a peak, then fallen into decline and finally collapsed. Yet our civilization is not destined to collapse, because we alone have experienced an industrial revolution.

The importance of this revolution cannot be overstated, because it gave rise to two new revolutionary classes. The class of people, of the middle ages, who were referred to as burgers, became the bourgeois, or capitalists, and they hired people to work for them, by the hour. This second class of people became known as the working class, or proletariat. I can only stress the fact that at first, both new classes, capitalist and proletariat, were revolutionary.

This changed dramatically, around the beginning of the twentieth century. At that time, capitalism reached the monopoly stage, otherwise known as imperialism. Monopoly capitalism has absolutely no progressive characteristics.

The monopoly capitalists, multi billionaires or imperialists, are completely reactionary. They are determined to run our society into the ground, in their never ending quest, for an ever greater profit.

In America, the richest country in the world, the multi billionaires never had it so good! They live in the lap of luxury! They pay no taxes!

By contrast, the working class is increasingly impoverished. Countless people are unemployed, homeless, hungry, living in cars or on the streets, under bridges and in subways. The food banks are running out of food. Alcohol and drug addictions are common place. Drug overdoses and suicides are frequent. Mass murders are now a daily event. Criminal gangs are in control of whole neighbourhoods. Drugs are being sold openly. The police are powerless.

As well, the small business owner, the middle class, or petty bourgeois, is being ruined, driven into bankruptcy, forced into the ranks of the working class.

This cannot go on. A revolution is necessary. The revolutionary working class, the proletariat, must overthrow the completely reactionary class of monopoly capitalists, the multi billionaires. At the same time, the existing state apparatus must be smashed, and replaced with a new state apparatus, in the form of the Dictatorship of the Proletariat.

It is the purpose of this little book, to outline precisely the manner in which this must be done.

INTRODUCTION TO THE FIRST EDITION

August 1, 2021. E Day. Eviction Day. The day that an estimated *eleven million* more Americans are scheduled to join the ranks of the homeless, at least if the landlords have their way. That is the day the moratorium on evictions expires, so that is the day the landlords plan to evict all the tenants who cannot pay their rent.

The wits among the working people respond to this news with that which passes for humour. "The good news it that the federal government has set aside $46 billion for rental assistance. The bad news is that only $3 billion is being spent."

The fact that the tenants are unemployed and also hungry, due in part to the Covid Virus, is of no concern to the landlords. But then the landlords are small time capitalists, middle class, "petty bourgeois". The buildings they own represent their invested capital, and they are determined to harvest the largest possible profit from their investment. That means collecting rent. Those who cannot pay the rent are free to live on the street. Nothing personal. Just business. Just capitalism.

Capitalism is precisely the problem! The capitalists are concerned with their profit, their "bottom line". Their goal in life is to make some "serious money", as they phrase it. The health and general well-being of the "common people", the "little guy", the "rank and file", the working class, the "masses" -a term I hate, as it sounds so impersonal- are of no concern to the capitalist. For precisely that reason, it is capitalism which has to be abolished and replaced with socialism.

This is not to say that the capitalists are indifferent, concerning the eviction of so many working class people. On the contrary, they are quite excited about this. They see this as an "opportunity", and the capitalists are constantly looking for opportunities. It is an opportunity to make a huge profit, so that many capitalists are now investing their capital in real estate. The evictions of so many working people can only work in favour of the capitalists!

The fact of the matter is that capitalism crushes and exploits all common people. Many such people are completely degraded, reduced to the level of beggars, relying on hand outs, merely to survive. All too often, they "self-medicate", through the use, and abuse, of alcohol and drugs. That is "one side of the coin", so to speak.

The other side of the coin is the fact that, every so often, people get into motion. "Serious motion", if you will excuse the expression. Revolutionary motion, to be precise. This is to say that millions of common people rise up and demand change. We are currently living in a time of revolutionary motion. People are fed up. There are limits! Countless people who were formerly apathetic are now "waking up", as they phrase it. Now is the time for change. Now is the time for socialism.

It is for the benefit of those who are just now "waking up", taking an interest in their lives, becoming politically active, demanding change, that I have chosen to write this little book. For that reason, certain important scientific terms are explained. Those who are already familiar with those terms may find this tiresome. Bear in mind that due to the revolutionary motion, countless working people are just now becoming politically active, and have to learn these terms. It is necessary to prepare them for scientific socialism. After all, it is the most advanced workers, the vanguard of the proletariat, who will lead the proletariat to socialism.

I say this because the one and only alternative to capitalism is "scientific socialism", in the form of the "Dictatorship of the Proletariat". Common people just have to be made aware of that.

CHAPTER 1

DEFINITION OF CLASSES

During a time of revolutionary motion, countless working people, those who were formerly apathetic, become politically active. They refer to this as "waking up". We are now living in a time of revolutionary motion, so that those "newly awakened", politically active working people, are questioning our political system, commonly referred to as "democracy".

Perhaps the first thing our freshly minted revolutionaries are going to have to learn, is that of the existence of classes. In North America, it is customary to deny the existence of classes, or at least that used to be the custom. No doubt the more advanced workers will check for a proper definition, on the internet. It is a wonderful invention, so why not use it?

With that in mind, our rising revolutionary star will learn the following: "Social classes are hierarchial groupings of individuals that are usually based on wealth, educational attainment, occupation, income or membership in a sub culture or social network." It goes on to say that "Many Americans recognize a three tier model that includes the upper class, the middle class, and lower or working class". In scientific jargon, we refer to this as the capitalist, or "bourgeois", definition of classes.

This stands in contrast to that which the internet defines as the "Marxist" definition of a class, which is defined as "a group with intrinsic tendencies and interests that differ from those of other groups within society".

That is not at all the "Marxist definition of class", but it gives common people a place to start. In fact, the monopoly capitalists, the billionaires, technically referred to as the "bourgeoisie", form one class. The small business owners, technically referred to as the "petty bourgeois", or middle class, form a second class. Those who have nothing to sell but their labour power, technically referred to as the "proletariat", or working class, form a third class. That leaves the family farmer, a fourth class, technically referred to as "peasants".

As there are very few family farmers left in North America, only the remnants of that class remain. Their influence, as a class, is negligible.

The internet also reports that most Americans believe that there are three classes in America, as previously listed. This is largely true, as the family farmers have been all but wiped out.

The belief in the existence of three classes in America, is a step forward from several years ago. At that time, the existence of classes was denied. Then, at around the time of the Occupy Movement, the working people began to sense the existence of classes. They began to refer to themselves as the "99 percent", as opposed to the "1 percent".

The understanding, at that time rather vague, was that the working people formed the vast majority of the population, the "99 percent", while the "1 percent", the "super rich", formed the tiny minority. There is some truth to this, and it was a step forward, on the road to class consciousness.

Since that time, the working class has advanced a little farther. Now it is widely understood that working people who are able to make ends meet are "middle class", the poverty stricken are "lower" or "working class", while the "super rich" or billionaires, are "upper class". Although still not completely accurate, it is a step forward from the understanding of

"99 percent" versus "1 percent". The latest "three tier model" at least acknowledges the existence of classes.

Much as I personally hate to refer to members of the working class, the proletariat, as "lower class", it is a term which is deeply entrenched in literature, as well as in common usage. In much the same way, I hate to refer to the "super rich", the monopoly capitalists, the multi billionaires, the "bourgeoisie", as members of the "upper class". Yet that expression also is deeply entrenched, so that there is nothing I can do about it. That merely leaves the family farmers, or "peasants", as well as the middle class, the small business owners, or "petty bourgeois".

Incidentally, the members of the "upper class", the "bourgeoisie", refer to themselves as "billionaires". The press also refer to them as billionaires. In the interest of not confusing the members of the working class, I too, have chosen to refer to them as billionaires, at least for the purposes of this article.

There are very few family farmers left in North America, and most of them are somewhat older. While previously, it was customary for the children to take over the family farm, that is no longer the case. Most young people are aware that there are "easier ways to go broke"! Family farming is "non stop work, with very little pay"! They simply cannot compete with the corporate agricultural monopolies! For that reason, the children of family farmers usually look for a job elsewhere. In other words, they become workers, proletarians. Who can blame them?

In fact, the vast majority of Americans are now working class, or "proletarians". That number is growing, as the family farmers are being ruined. The members of that class are being forced into the ranks of the proletariat.

That leaves only the members of the middle class, technically referred to as "petty bourgeois". These people tend to be owners of small businesses, in that such a business is not part of a corporate monopoly. By and large, they are living on "borrowed time", as the corporations have yet to drive them into bankruptcy. Such businesses could include the corner store, or a laundromat, for example. All are about to go the way of the dodo bird. The owners of those businesses too, are about to join the proletariat.

The monopoly capitalists, the billionaires, working through their corporations, are doing a fine job of eliminating all competition, regardless of how slight!

To be clear, the members of the working class, the proletarians, have nothing to sell but their labour power. They sell themselves, by the hour, because they have nothing else to sell!

By contrast, the people who own all the factories, mills, mines, railroads, airlines, shipping lines and internet, as well as everything else of any considerable value, are referred to as capitalists. Not that they refer to themselves as such. They prefer to refer to themselves as "entrepreneurs". As if changing the name, changes the nature of the beast!

The billionaires, monopoly capitalists, make their profit from the labour of their workers, so it is in their interests to pay their workers as little as possible, while forcing them to work as hard as possible. It is in the interest of the workers to sell themselves for the highest possible price. This is to say that the interests of the capitalists, the billionaires, and the interests of the workers, the proletarians, are "diametrically opposed". That which is in the best interest of one class, is in the worst interest of the other class.

These scientific terms I have placed in quotation marks, as it is important that all working people should learn their meaning. Otherwise, the capitalists will not hesitate to use our lack of awareness of these terms, against us.

I should add that, for the purposes of this article, I refer to the billionaires, and only the billionaires, as capitalists. Also, allow me to stress the fact that I have no quarrel with the small business owners. They are being ruined, forced into the ranks of the working class. Welcome, my Brothers and Sisters, my Comrades!

To such people, I can only say that you have no future under capitalism. The billionaires have clearly stated that they fully intend to ruin all small businesses. Take them at their word. They mean it!

You can believe the billionaires when they say that *all* banks and businesses, with assets of *less than a Trillion,* are *Too Small To Succeed!*

Three banks have already failed. Numerous others are "tottering on the brink". Almost all of them are about to fail, which will also bring down countless "small businesses". This will give rise to a Second Great Depression. *Unless the billionaires are stopped!*

The working class has been "bled white", so to speak. There is not much more they can squeeze out of the workers! To quote that old expression, "You cannot get blood out of a stone"! So now the billionaires are turning their attention to the middle class, the small business owners! It is simply a matter of driving those businesses into bankruptcy, and then "seizing all the assets", as they phrase it. As some of these "small businesses" have "assets" of many billions, there is clearly a "rich crop" to be harvested!

It is best to give credit where credit is due! The *billionaires* are both *ambitious* and *focused!* They are *focused* on becoming *Trillionaires!*

If nothing else, this simplifies the class struggle, especially in America, which is the name by which all citizens of the United States refer to their country. They also refer to themselves as Americans. Out of respect for those people, I have chosen to use the same terms.

As is well known, Americans have long since given the nobility, otherwise known as the monarchy, their "walking papers". Properly so! The King of England is of no concern to Americans. In America, the class of people known as the nobility, does not exist.

This is another way of saying that the class struggle is now sharp and clear. Very simple. Us against them. The workers versus the capitalists. The proletariat battling the billionaires. No holds barred. No quarter. Winner takes it all. War to the finish.

CHAPTER 2

DEMOCRACY: A METHOD OF CLASS RULE

Democracy was first created in the Greek city-state of Athens, many years ago. At that time, there were basically two main classes, slaves and slave owners. Yet the slave owners devised a method of rule, in which the slave owners voted on all affairs of state. Majority rule! Democracy! Of course, the slaves had nothing to say about this!

In fact, the slaves had a rather "annoying" habit of rebelling! For that reason, it was necessary for the slave owners to devise a system, a "state apparatus", in order to keep the slaves "in their place". With that in mind, the slave owners created a force of armed men. It was their duty to hunt down and capture any escaped slaves, and to crush any slave uprising.

This is to say that the earliest form of democracy, was indeed a democracy, but *only* for one class, the class of slave owners. For the slaves, democracy was a dictatorship! That dictatorship was enforced, through a *state apparatus*, in the form of a body of armed men, with violence!

In scientific jargon, we say that democracy has a "dual nature". It is *both* a democracy, and a dictatorship! For the class of people in charge, it is a democracy, but for the class of people being crushed, it is a dictatorship.

Over the years, as society has evolved, different classes have come into existence, and different forms of state apparatus have been created. Yet all forms of state apparatus serve the same purpose. It is the method by which one class, the ruling class, uses to suppress all subordinate classes. That is true to this day.

In America, the capitalists, the billionaires, are the ruling class. The working class, the proletariat, is the class being ruled. The method of rule, of the billionaires, is that of the democratic republic. Democracy for the billionaires, but a dictatorship for the working class.

As anyone who has recently watched the news can testify, the journalists are focused on the "events of January 6", which they refer to as the "one six insurrection". As I have previously documented, it is certainly true that a window was broken, and clearly a police officer was knocked to the ground. If that is their idea of an insurrection, then they are about to receive a rude awakening!

Of course, the capitalists are terrified, as on that day, a great many peaceful protesters -along with a few vandals- marched on the capitol, walked into the government building, almost unopposed, and terrified the politicians. In the process, they exposed the weakness of the capitol.

Of course the journalists are giving their own "spin" to the story. At first, the protesters were accused of being "anarchists", in the service of Donald Trump, as they are convinced that "Trump won the election". The journalists refer to this as the "Big Lie". They could not make that "stick", as it was pointed out to them that anarchists are determined to have no government. As the protesters were determined to reinstate Trump as president, they could not possibly be anarchists.

So now the journalists have "changed their tune". They are now saying that the "one six insurrection" was a "threat to democracy". This is closer to the truth. In fact, it is a threat to *their* democracy, the democracy of the *billionaires*, "bourgeois democracy". Such little details the journalists neglect to mention!

I am sure our "newly awakened", politically active members of the working class are skeptical. After all, it is commonly thought that democracy is "majority rule". Or is it? Here too, such people will check with the internet.

In fact, the internet provides a definition of democracy, which is quite close: "A system of government by the whole population or all the eligible members of a state, typically through elected representatives". That is very close to the definition of democracy which is taught in schools, that of "majority rule".

The Marxist understanding of democracy is far different! Lenin explains this quite well in his book, State and Revolution: "In capitalist society, under the conditions most favourable to its development, we have more or less complete democracy in the democratic republic. But this democracy is always restricted by the narrow framework of capitalist exploitation, and consequently always remains, in reality, a democracy for the minority, only for the possessing classes, only for the rich. Freedom in capitalist society always remains about the same as it was in the ancient Greek republics: freedom for the slave owners. Owing to the conditions of capitalist exploitation, the modern wage slaves are also so crushed by want and poverty that 'they cannot be bothered with democracy', 'they cannot be bothered with politics'; in the ordinary peaceful course of events the majority of the population is debarred from participating in social and political life."

That is, without doubt, a more accurate, detailed description of democracy, a far cry from the "majority rule" of popular belief. The discrepancy can be explained by the fact that the billionaires own the internet, just as they own all the major news outlets, so that all information, including breaking news, is biased in favour of the capitalists.

Yet the fact remains that under the "democratic republic", we have "more or less complete democracy". This democracy is "restricted by the narrow framework of capitalist exploitation", so that it remains a democracy "for the rich". We refer to this democracy, under capitalism, as "bourgeois democracy". It is a *democracy* for the billionaires, but a *dictatorship* for the working class.

It should also be pointed out that there is a big difference between a republic, and a country which recognizes a monarch. A republic does not recognize any monarch, which is to say a king or queen, so that the United States is a democratic republic. By contrast, Canada does recognize the King of England as the Head of State, the King of Canada. Canada is a Constitutional Monarchy.

As I have documented, in a previous article, some of the ways in which democracy is curtailed in America, there is no need to repeat it here. Suffice it to say that it is the members of the Electoral College who elect the President and the Vice President, not the voters. The popular vote is a mere formality.

Democracy is also curtailed in Canada, only in a different manner.

Upon the death of Queen Elizabeth, it was her son, who ascended to the throne. He is now King Charles, the Canadian Head of State. He is in turn represented in Canada by a Governor General, in this case Mary Simon. It is Mary Simon whom in turn chooses the Canadian head of government, in our case the Prime Minister. She also has the power to act as a restraint on the power of the Prime Minister. According to the internet -a most valuable source of information, however biased- her responsibilities include "carrying out constitutional duties, serving as Commander in Chief, representing Canada at home and abroad, encouraging excellence, and bringing Canadians together".

Perhaps there is some confusion concerning the "state", as opposed to the "government". Most people consider them to be the same, different names for one entity. But as His Majesty is the King of Canada, our Head of State, but not part of our government, the state and government, cannot be the same.

In order to provide some clarity, I checked with the internet. They provided their own definition: "The state is the organization, while the government is the particular group of people, the administrative bureaucracy that controls the state apparatus at a given time. That is, governments are the means through which state power is employed". Now we know.

In the case of Canada, the state organization, or "machine", has been set up by the nobility and the capitalists, the billionaires, in order to crush and exploit the common people, the working class. The precise individuals who serve in any capacity, perhaps as Prime Minister, at any given time, may vary from one year to the next, or at best from one election to the next. They form the "government", but regardless of the name of the political party, all serve the same class or classes, in our case the nobility and the billionaires.

As an example, let us assume that the Conservatives have a majority in Parliament, so that the Prime Minister is a member of the Conservative Party. Let us further assume that at the time of a federal election, a different political party, perhaps the Liberals, manages to achieve a majority in Parliament.

At that point, it is the duty of the Governor General to appoint a new Prime Minister, a member of the Liberal Party, as the new Prime Minister.

The new Prime Minister could then be expected to appoint her cronies, "party faithful", to key positions within the "new government". This is referred to as the "bureaucracy", in that these "government officials" are appointed, not elected. This includes members of the Cabinet, for example. Different government, different faces, same state apparatus, serving the same capitalist class, crushing the same working class.

As Marx phrased it, federal elections allow the working class to "decide once in every three or six years which member of the ruling class was to misrepresent the people in Parliament". Which in no way lessens the importance of the federal elections. It stands as an "index of maturity of the working class", according to Marx. For that reason, if for no other, it is of vital importance to vote in all elections.

Here we have another fine example of democracy which is "restricted". Just as Americans do not elect their President, so too Canadians do not elect their Prime Minister. That little detail is covered by the Governor General, the same lady whose job it is to "bring us together". Now that is a tall order! If there is any group of people who are more divided than the Americans, it is the Canadians!

This is significant, because it is just a matter of time before the revolution breaks out, possibly first in America. Then it will almost certainly spread to Canada. The two countries have a great deal in common. It is very likely that both countries will break up into separate independent socialist republics, so that Canadians will finally kiss the British monarchy good bye! It cannot happen too soon!

It is also quite possible that areas of both countries will unite. That remains to be seen, and is completely up to the people who live in those areas. Yet two examples come immediately to mind. The fact is that northern New England has a great deal in common with the Maritime Provinces, just as Alaska has a great deal in common with the Yukon Territory. The idea that those areas could merge, into separate socialist republics, is quite conceivable.

C H A P T E R 3

SCIENTIFIC SOCIALISM

We have just established the fact that now, thanks to the billionaires, only two main classes remain in existence. Those are the working class, the proletariat, and the capitalist class, the billionaires. As well, there is a war raging between them. This war has been raging for perhaps three hundred years, ever since the industrial revolution gave birth to those two classes. After all, one class cannot exist without the class opposite to it, or "antipode", as is the correct scientific term. Sometimes this war in hidden, "simmering", as it were, and sometimes it "flares up" into open rebellion. We are very close to the point that the class war will explode into armed conflict. We had best be prepared. This is to say that we had best focus on our goals.

This calls for a little explanation. Some people may find such details tiresome, but bear with me, as it is important. The working people who are just now becoming politically active, must become aware of this. Or to put it another way, they must be "brought up to speed".

At the time of the industrial revolution, which started in Great Britain, the class of people known as the burghers, saw an opportunity to invest in factories, mills, mines, railroads, shipping lines and such, and make considerable money. They succeeded, beyond their wildest dreams. In the process, they became known as "bourgeois", a corruption of the word burgher. They in turn, were forced to hire people to work for them, mainly peasants and artisans. These people were transformed into workers, or "proletarians", so that two new classes were created.

The capitalists, or bourgeois, knew that they were making a huge profit. They did not know *how* this was happening, and they really did not care. That is largely true to this day! Their one and only concern is with their profit, their "bottom line", as they phrase it.

Yet the bourgeois economists of the day were concerned, but could not figure it out.

Karl Marx explained it! But then, Karl Marx was not a "bourgeois" economist!

It was Marx who first conducted a *scientific* examination of capitalism. In the process, it was Marx who first explained how the capitalists make their profit. More than that, he determined a few other things, but perhaps it would be best to allow Marx to explain this, in his own words.

In a letter to Weydemeyer, dated March 5, 1852, he stated: "And now as to myself, no credit is due to me for discovering the existence of classes in modern society, nor yet the struggle between them. Long before me, bourgeois historians had described the historical development of this class struggle, and bourgeois economists the economic anatomy of the classes. What I did that was new was to prove: 1) that the *existence of classes* is only bound up with *particular historical phases in the development of production;* 2) that the class struggle necessarily leads to the *Dictatorship of the Proletariat;* 3) that this dictatorship itself only constitutes the transition to the *abolition of all classes and to a classless society.*"

This letter of Marx is of the utmost importance, if for no other reason that it allows us to distinguish the true Marxists, from the social chauvinists. Such people are those who are Marxists in name only, chauvinists in deeds.

By this I mean that a social chauvinists is someone who claims to be a Marxist, while denying the necessity of the Dictatorship of the Proletariat. They are of the opinion that the theories of Marx and Lenin must be revised. As can be well imagined, this is referred to as "revisionist".

Incidentally, I have chosen to place capitols on the Dictatorship of the Proletariat. That is precisely what Engels did, in 1891, in his introduction to the article by Marx, titled Civil War In France: The Paris Commune. He clearly considered this to be of the utmost importance!

Lenin was also of this opinion, as in his book, State And Revolution, he stated. "The theory of the class struggle was *not* created by Marx, but by the bourgeoisie *before* Marx, and generally speaking is *acceptable* to the bourgeoisie. Those who recognize *only* the class struggle are not yet Marxists; these may be found to have gone no further than the boundaries of bourgeois reasoning and bourgeois politics. To limit Marxism to the theory of the class struggle means curtailing Marxism, distorting it, reducing it to something which is acceptable to the bourgeoisie. A Marxist is one who *extends* the acceptance of the class struggle to the acceptance of the Dictatorship of the Proletariat. This is where the profound difference lies between a Marxist and an ordinary petty (and even big) bourgeois. This is the touchstone on which the *real* understanding and acceptance of Marxism should be tested." (italics by Lenin)

It is clear that Lenin was a true *social scientist*, one who followed in the footsteps of Marx and Engels. They knew that society develops according to certain *laws*, and that capitalism, of necessity, *develops* into *socialism*, in the form of the Dictatorship of the Proletariat.

In fact, after years of careful study, in 1848, Marx and Engels wrote the Communist Manifesto. It gave the working class people of all countries a proper direction. Yet they had no idea of the *form* the new working class government should take, under socialism, as they had nothing upon which to base it.

It was not until 1871, that the class struggle, between the workers and the capitalists, broke out into open warfare. In that year, the workers of Paris revolted and set up a working class, socialist government. They referred to this as the Paris Commune. They referred to themselves as the Communards. In honour of the Paris Communards, those of us who are Marxists, now refer to ourselves as Communists.

The Paris Commune was brief, lasting a mere few weeks, before it was crushed, with the utmost brutality. Yet it provided Marx with the information he needed. He referred to this as the "positive form of a republic that was not only to supersede the monarchial form of class rule, but class rule itself".

In 1871, Marx was living in Britain, but was closely following the developments in France. As the revolt in Paris unfolded, he advised the Communards on a proper course of action. Advice they did not follow, to their everlasting regret.

One of the biggest mistakes the Communards made, was in *not destroying* the existing state apparatus, that which has been set up by the capitalists, in order to *crush* the working class. Not that the Communards can be faulted for this. They just did not know any better.

Yet the Paris Commune, brief as it was, provided Marx with the information he required. But perhaps it is best to allow Lenin to describe this, as he stated in State and Revolution:

"Marx, however, was not only enthusiastic about the heroism of the Communards who 'stormed the heavens', as he expressed it. Although it did not achieve its aim, he regarded the mass revolutionary movement as a historic experiment of gigantic proportions, as an advance of the world proletarian revolution, as a practical step that was more important than hundreds of programs and discussions. Marx conceived his task to be to analyze this experiment, to draw lessons in tactics from it, to reexamine his theory in the light it afforded.

"Marx made the only 'correction' he thought it necessary to make in the Communist Manifesto on the basis of the revolutionary experience of the Paris Communards.

"The last preface to the new German edition of The Communist Manifesto, signed by both its authors, is dated June 24, 1872. In this preface the authors, Karl Marx and Frederich Engels, say that the program of the Communist Manifesto 'has in some details become antiquated' now, and they go on to say: 'One thing especially was proved by the Commune, viz., *that the working class cannot simply lay hold of the ready-made state machinery and wield it for its own purposes!*'" (italics by Lenin)

Lenin went on to say, "if the state is the product of irreconcilable class antagonisms, it is a power standing *above society* and *'increasingly alienating itself from it'*, it is clear that the liberation of the oppressed class is impossible, not only without violent revolution, *but also without the destruction* of the apparatus of state power which was created by the ruling class and which is the embodiment of this 'alienation'". (italics by Lenin)

That last sentence spells out quite well, the point of State and Revolution! In order to "liberate the oppressed class", in our case the working class, a "violent revolution" is necessary! At the time of this "violent revolution", the "apparatus of state power", must be destroyed!

This is to stress the fact that Lenin was every bit the social scientist, as was Marx! Lenin was determined to *not* repeat the mistakes of previous revolutions, as was pointed out by Marx!

Lenin wrote State and Revolution in early 1917, in anticipation of the approaching Russian Socialist Revolution. The common people of Russia, the workers and peasants, followed his advice. For that reason -and only for that reason!- the Russian Revolution, of October 25, old style calendar, or November 7, new style calendar, was successful. This has gone down in history as the Russian Great October Socialist Revolution.

The Russian capitalists and landlords, as represented by Kerensky and his Provisional Government, was overthrown, and a Soviet Socialist Republic was established. As most of the common people of Russia were peasants, it became known as the Dictatorship of the Proletariat and Poor Peasants.

The experience of the Russian Great October Socialist Revolution, provided Lenin with a wealth of information. As the true scientific socialist that he was, after three years of socialism, under the Dictatorship of the Proletariat, he considered it to be his *duty* to analyze the experience that was gained.

For that reason, he wrote another article, in the spring of 1920. It is a true master piece, titled Left Wing Communism, An Infantile Disorder.

As he stated, "We now possess quite considerable international experience, which shows very definitely that certain fundamental features of our revolution have a significance that is not local, or particularly national, or Russian alone, but international….not merely several but all the primary features of our revolution, and many of its secondary features, are of international significance in the meaning of its effect on all countries… taking international significance to mean the international validity or the historical inevitability of a repetition, on an international scale, of what has taken place in our country. It must be admitted that certain fundamental features of our revolution do possess that significance".

This is to say that the approaching American Revolution is an "historical inevitability", and is about to become a "repetition", of the Russian Great October Socialist Revolution! Just as the "absolute centralization and rigorous discipline in the proletariat", was an "essential condition of victory", of the success over the Russian capitalists, so too, that is also necessary in America!

At the time Lenin was writing this, the country of Russia had a strong "proletarian revolutionary party". Bear in mind that *at the moment*, America is not blessed with such a party. At least, not yet! But in Russia, at that time, the discipline of the party was "tested" and "reinforced" by the "class consciousness of the proletarian vanguard and by its devotion to the revolution, by its tenacity, self-sacrifice and heroism. Secondly, by its ability to link up, maintain the closest contact, and -if you wish-merge, in certain measure, with the broadest masses of the working people- primarily with the proletariat, *but also with the non-proletarian* masses of the working people. Third, by the correctness of the political leadership exercised by this vanguard, by the correctness of its political strategy and tactics, provided the broad masses have seen, *from their own experience,* that they are correct...Their creation is facilitated by a correct revolutionary theory, which, in its turn, is not a dogma, but assumes final shape only in connection with the practical activity of a truly mass and truly revolutionary movement." (italics by Lenin)

It is important to bear in mind that Lenin here stresses the *importance* of the "class consciousness of the proletarian vanguard", by whom he means the most advanced workers. It is supremely important to raise their level of awareness! At the same time, the "broad masses", which in our case means the less advanced workers, have to learn, *"from their own experience",* that the most advanced workers, those who are aware of the revolutionary theories of Marx and Lenin, are correct.

This brings us to the definition of the word "dogma". According to the internet, it is "a principle or set of principles laid down by an authority as incontrovertibly true". That is *not* Marxism! As Lenin stated, "revolutionary theory is not a dogma"! It is a guide to action!

Here is another statement of Lenin, which is of the utmost importance, under the present circumstances:

"The fundamental law of revolution, which has been confirmed by all revolutions and especially by all three Russian revolutions in the twentieth century, is as follows: for a revolution to take place it is not enough for the exploited and oppressed masses to realize the impossibility of living in the old way, and demand changes; for a revolution to take place it is essential that the exploiters should not be able to live and rule in the old way. It is only when the *'lower classes' do not want* to live in the old way and the 'upper classes' *cannot carry on in the old way* that the revolution can triumph. This truth can be expressed in other words: revolution is impossible without a nationwide crisis (affecting both the exploited and the exploiters). It follows that, for a revolution to take place, it is essential, first, that a majority of the workers (or at least a majority of the class conscious, thinking, and politically active workers) should fully realize that revolution is necessary, and that they should be prepared to die for it; second, that the ruling classes should be going through a governmental crisis, which draws even the most backward masses into politics (symptomatic of any genuine revolution is a rapid, tenfold and even hundredfold increase in the size of the working and oppressed masses -hitherto apathetic- who are capable of waging the political struggle), weakens the government, and makes it possible for the revolutionaries to rapidly overthrow it". (italics by Lenin)

In my opinion, that statement is so important, it should be created in poster form, and posted in all union halls, and the homes of all working class people!

The conditions which Lenin described, which are necessary for a revolution to take place, now exist in North America, as well as in other countries of the world. It is also a fact that Soviets, otherwise known as Councils, first appeared in Russia, at the time of the 1905 Revolution. We now know that these Soviets are a spontaneous creation, of the working class, during times of revolutionary motion. Those Soviets are now also appearing again, in North America.

Now all working class people must become aware of the *meaning*, and the *necessity* of, the terms Soviet Power and the Dictatorship of the Proletariat. This is referred to as "raising the level of awareness of the proletariat"!

In my opinion, this is now of critical importance! This is now the "key link" in the current "chain of events", which are certain to lead to the Second American Revolution! The working class must now learn the meaning of, and *embrace, Soviet Power* and the *Dictatorship of the Proletariat!*

CHAPTER 4

DICTATORSHIP OF THE PROLETARIAT

After the forthcoming proletarian revolution, the billionaires will be overthrown, the state apparatus will be smashed, and a new state apparatus will be established, in the form of the Dictatorship of the Proletariat. This is referred to as scientific socialism.

Yet classes will continue to exist! The billionaires will be "down but not out"! They are not about to "resign themselves to their fate"! Such a thought would never cross their minds! On the contrary, they will be determined to "regain their paradise lost"! They will stoop to any depth, any lie, any subterfuge, in order to return to power. For that reason, they must be crushed! That requires a state apparatus!

As for those who are skeptical, bear in mind that no one wants to *lower their standard of living!* That includes the billionaires! They are accustomed to "living in the lap of luxury"! Servants wait on them, "hand and foot"! They live in houses which can best be described as *mansions!* They have *fleets of luxury vehicles!* Also *yachts!* And *jet aircraft,* to fly them anywhere in the world! They are also accustomed to *paying no taxes!* They are *not* about to *embrace socialism!*

For that reason, they must be crushed!

Yet the state apparatus which was set up by the capitalists, in order to crush the working class, cannot be used. It must be destroyed. So that requires a different state apparatus, in order to crush the "desperate and determined resistance", of the billionaires, as they try to regain their "paradise lost". That new, proletarian state apparatus, is referred to as the Dictatorship of the Proletariat.

It is vitally important to understand this, especially as the American revolution could break out, at any time. The American revolutionaries must not repeat the mistakes of the heroic revolutionary workers of Paris, those who took part in the Paris Commune! The biggest mistake they made was in *not* smashing the existing state machine! Instead, they tried to *take over* the existing state machine, and use it for their own purposes! A huge mistake!

The existing state machine has been carefully designed by the capitalists, for the express purpose of *crushing* the working class. It stands to reason that the people who manage to take control of that state machine, at the time of the revolution, will inevitably end up using it, for that *same purpose!*

This is to stress the fact that the existing state machine must be *smashed!* Otherwise, the people taking part in the revolution, will *take over* the state machine, and set themselves up as the *new rulers!* The revolution will succeed only in replacing one set of rulers with another! Out of the frying pan, into the fire! Not a vast improvement!

We can expect the social chauvinists, those who merely *claim* to be Marxists, to bitterly *oppose* the *destruction* of the state apparatus! Well of course! Their plan is to take control of that state apparatus, at the time of the revolution, and set themselves up as the new rulers! There is a method to their madness! And people wonder why true Marxists cannot get along with them!

The important thing to remember is that the political parties which claim to be Marxist, but *deny* the necessity of smashing the existing state apparatus, and replacing it with the Dictatorship of the Proletariat, are *revisionists*, "social chauvinists", socialists in words, chauvinists in deeds! Their *goal* is to take over the existing state apparatus, at the time of the revolution, and *set themselves up as the new rulers!*

They are defending the billionaires, not to be trusted. We *cannot* work with them, although there is hope for the rank and file members. The leaders of such parties are almost certainly well *aware* of the revolutionary theories of Marx and Lenin, and equally determined to *distort* those theories. They are devoted servants of the billionaires.

As they also have the ideology of the capitalist, they see the revolution as an "opportunity". It is simply a matter of pretending to be a Marxist, establishing themselves as leaders of the working class, and at the time of the revolution, taking over the existing state apparatus. At that point, they can set themselves up as the new rulers, perhaps even as the first "Marxist" President! We already had a Black President, so why not a Marxist President?

Yet if the working people, those who are taking part in the revolution, manage to mount a successful insurrection, storm the capitol and smash the existing state apparatus -God forbid!- then the juicy political plums, such as the presidency, will be lost forever! For that reason, the working people must *not* be made aware of the revolutionary theories of Marx and Lenin. Logic of the social chauvinists! There is a method to their madness!

They would have us believe that after the revolution, all workers and -former!- billionaires can unite as brothers. They would have us embrace the fable of "the lion shall lie down with the lamb". Mankind will then enter an "Age of Aquarius", in that "peace and tranquility, equality and understanding" shall reign supreme. Fairy tale! That is not about to happen! It may be written in the stars, according to astrologers, but nowhere is it engraved in stone, here on planet earth! Even if it is engraved in stone, it remains a barefaced lie!

Feel free to face reality! The fact is that it is possible to *defeat* the billionaires at one stroke, possibly by a successful uprising in the capitol, similar to that which happened on January 6, or "one six", as is the common expression. In such a manner the capitalists will be *overthrown*, but not *destroyed!* Even after the capitalists are deposed, they will still remain *stronger than the proletariat!* Facts! Unless they are suppressed, absolutely crushed, *after* they are overthrown, they will most certainly *return to power!*

For that reason, *after* the revolution, the -former!- billionaires, must be *crushed!* They must be allowed no opportunity to return to power! We are not anarchists. We are well aware that even *after* the revolution, *after* the billionaires are overthrown, *after* the existing state apparatus is destroyed, *classes will still exist!* A *different* state apparatus is required, in order to prevent the billionaires from returning to power! That state apparatus is referred to as the *Dictatorship of the Proletariat!*

In fact, Lenin expressed himself quite clearly on this very point: "Dictatorship is rule based directly upon force and unrestricted by any law. The revolutionary Dictatorship of the Proletariat is rule won and maintained by the use of violence by the proletariat against the bourgeoisie, rule that is unrestricted by any law...dictatorship presupposes and implies a 'condition', one so disagreeable to renegades, of *revolutionary violence* of one class against another".

As Lenin stated, after the revolution, "the organ of suppression is now the majority of the population, and not the minority, as was always the case under slavery, serfdom and wage-slavery. And since the majority of the people *itself* suppresses its oppressors, a 'special force' for suppression is *no longer necessary*". (italics by Lenin)

The point must be stressed, to the working class, that the existing state apparatus, which has been set up by the billionaires, with the express purpose of crushing and exploiting the vast majority of working people, must be *smashed*. This state apparatus includes the police, the standing army, the bureaucracy, the prisons and other coercive institutions. It must be *replaced* by a different state apparatus, a proletarian state apparatus, with the goal of crushing the desperate and determined resistance of the billionaires, as they try to return to power.

We can expect the social chauvinists -the traitors to socialism!- to "water this down", to refer to it as the "rule of the working class". It is absolutely *not* a "rule"! It is a *Dictatorship!* Under the Dictatorship of the Proletariat, the billionaires will have *no rights*! As well, the criminal elements, the thieves and killers, the mobsters, the sex offenders, the rapists and child molesters, the drug dealers, will have *no rights*! They will *not* be allowed to hide behind lawyers! There will be *no* statute of limitations! They will *not* be molly coddled! They *will* be held accountable! Those who are convicted, will *not* be sent to a country club of a prison! They *will* be put to work! They *will* be forced to perform useful, productive work, for perhaps the first time in their lives!

Bear in mind that Lenin refers to the Dictatorship of the Proletariat as the "very essence of proletarian revolution". He goes on to say that, "This is a question that is of the greatest importance for all countries, especially for the advanced ones, especially for those at war, and especially at the present time. One may say, without fear of exaggeration, that this is the key problem of the entire proletarian class struggle. It is therefore, necessary to pay particular attention to it."

As the Dictatorship of the Proletariat is of such "great importance", we can expect the capitalists to attack it, most savagely. After all, it is the greatest nightmare of all billionaires. With good reason, I might add!

We can expect the modern day journalists, devoted servants of the billionaires, one and all, to compare the two "methods of rule", that of the "democratic", to the "dictatorial". Nonsense! We must stress, at every opportunity, that democracy is a method of *class rule!* Either the democracy of the billionaires, which is a dictatorship over the working class, or a democracy of the working class, which is a dictatorship over the billionaires, the Dictatorship of the Proletariat.

No doubt the journalists will sing the praises of the social chauvinists, the Benedict Arnolds of Marxism. They will be praised as the "more moderate elements", "true democrats", as opposed to those who call for the Dictatorship of the Proletariat.

Our response, in every possible situation, must be to stress the fact that democracy is a *method of class rule!* We must constantly raise the question: *Democracy for which class?*

Further, proletarian democracy is far superior to bourgeois democracy. The one and only form of proletarian democracy lies in the Dictatorship of the Proletariat. That is democracy *of* the working class, *by* the working class, *for* the working class.

At every rally, every demonstration, every political gathering, our posters and banners must call for the Dictatorship of the Proletariat! It is in this manner, and *only* in this manner, that the level of awareness of the working class can be raised! They will also become aware of the fact that the Dictatorship of the Proletariat is the very *essence of the doctrine of Marx!*

This is necessary, as very few working class people have been to university, so they have not been exposed to the revolutionary theories of Marx and Lenin. Bear in mind that it is only in university, that those revolutionary theories are taught. Or more accurately, those theories are *distorted* in university.

The social chauvinists may also insist that a dictatorship implies the rule of a single person. It most certainly does *not*, so it is up to the true Marxists to draw the attention to the existence of classes. *Class rule!*

In desperation, they may even resort to accusing the Marxists of threatening to use *force* to stay in power, as dictatorship implies a rule unrestricted by any laws. This is most emphatically *true!* In fact, Lenin expressed himself quite clearly on this very point: "Dictatorship is rule based directly upon force and unrestricted by any law. The revolutionary Dictatorship of the Proletariat is rule won and maintained by the use of violence by the proletariat against the bourgeoisie, rule that is unrestricted by any law...dictatorship presupposes and implies a 'condition', one so disagreeable to renegades, of *revolutionary violence* of one class against another". (italics by Lenin)

Such a state apparatus must be different from the bourgeois state apparatus, that which was set up by the capitalists, in order to crush the working class. In the case of Russia, it was destroyed at the time of the Great October Socialist Revolution.

As Lenin described it, after the revolution, "the organ of suppression is now the majority of the population, and not the minority, as was always the case under slavery, serfdom and wage-slavery. And since the majority of the people *itself* suppresses its oppressors, a 'special force' for suppression is *no longer necessary*". (italics by Lenin)

Another popular argument of the social chauvinists, the traitors to Marxism, is that as we have such a huge majority, far more workers than capitalists, then there is no need to crush the billionaires, after the revolution. So why do we need a dictatorship?

Because without the Dictatorship of the Proletariat, after the revolution, the billionaires are certain to *return to power!*

This is not to say that the billionaires are worried about being overthrown. On the contrary, at the moment, meaning just before the revolution, the billionaires are not even concerned! It is characteristic of any and all ruling classes, and the billionaires are no exception, to believe that their rule will last forever! It just never occurs to them that they can be overthrown! Such thoughts never cross their mind! And no wonder! They pay their flunkies to worry about such little details!

These flunkies, belly crawling boot lickers, one and all, are paid quite handsomely, to tell the billionaires that which the billionaires want to hear! They earn their pay! They assure the billionaires that "all bases are covered", "there is nothing to worry about", "everything is under control", the sounds from the streets are merely "loud noises". The billionaires believe them!

In fact, the "loud noises" from the streets, are the sounds of revolution! The protests, the marches and demonstrations, are the sound of the working class rising up! The Autonomous Zones, which are springing up across the country, are nothing other than Soviets, Councils of workers!

The states which are uniting, in various parts of the country, are about to declare independence and form separate socialist republics! At the time of the revolution, they will also tell Washington what they can do with their national debt!

It is important that we draw a clear distinction between the social chauvinist, and the utopian socialists. The utopian socialists make no claim to be Marxists. They are people of principle! We can certainly work with them!

It matters not if they refer to themselves as Independent Socialists, or Social Democrats, or Democratic Socialists or just plain Socialists. They are fighting for socialism, or at least for reforms under capitalism. Bear in mind that reforms "strengthen and further the revolutionary movement", according to Marx, so that we can absolutely work with them. But then again, we must still be allowed to put forward our own belief in revolution, smashing the existing state apparatus, and establishing the subsequent Dictatorship of the Proletariat. That is a necessary condition for our involvement with them.

No doubt the members of the working class who have just recently "woken up", in that they have become politically active, are bound to be confused by the numerous "Leftist" organizations and political parties. Who can blame them? Some of these organizations claim to be Marxists, others say they are Independent Socialists, or Democratic Socialists, or just plain Socialists. Some of them claim to be parties, while others make no such claims. It does not help that some of those who claim to be Socialists, maintain that while socialism is a good idea, it is simply not possible!

To such freshly awakened revolutionaries, I can only state that the social chauvinists are to be avoided! Those who claim to be Marxists, while maintaining that the theories of Marx and Lenin, must be "revised", are completely devoid of principle! Opportunists! They are dead set *opposed* to Soviet Power and the Dictatorship of the Proletariat! Devoted servants of the billionaires!

The only true Marxists, Communists, are those who call for revolution, Soviet (Council) Power, the destruction of the existing state apparatus, and the subsequent Dictatorship of the Proletariat.

That is the *one and only* way in which the billionaires can be overthrown, and socialism established and maintained! *True Scientific Socialism!*

In the interests of driving home the point that the scientific socialism, of Marx and Lenin, has been proven to be the *only* true form of socialism, allow me to give a few examples.

During the time that Marx worked, there were several individuals, utopian socialists, fine fellows, who tried to create socialism, in their own way, under capitalism.

Perhaps the most famous, at least in Great Britain, was Robert Owen, a British industrialist, who lived from 1771 to 1858. He truly cared about people, especially the folks who worked for him. He believed, as all utopian socialists believe, that the best way to enact change, is by first "changing moral values and external conditions". With that in mind, he set up a textile manufacturing plant in a place referred to as New Lenark, in Scotland. His goal was to create a community in which "workers were paid what they were worth, and shared everything".

No one can find fault with his ambition. He made a supreme effort. And failed miserably!

There were several other individuals, also utopian socialists, who conducted similar experiments in Europe, at around the same time. All of these experiments ended in failure, through no fault of the utopian socialists. It is simply not possible to create socialism, under capitalism!

This has not stopped other utopian socialists from attempting to create similar socialist societies. Senator Bernie Sanders of Vermont, who describes himself as an Independent Socialist, is perhaps the most famous, and certainly well respected. He is also just as mistaken as Owen. In time, he and his followers will find that we can live under capitalism, or socialism, but not both, not at the same time.

CHAPTER 5

CLASS CONSCIOUSNESS

I t is one thing to say that the working class is cultured, as is certainly the case. It is something else entirely to say that the working class is *class conscious*, as it most certainly is *not!*

The conditions of life, of the working class, does not lead to the awareness of itself, as a *class!*Nor does it lead to the awareness of the *class conflict!* Yet this in no way changes the fact that working people get into motion, frequently by the millions, and *make history! Revolution! Made by people who are not conscious of their actions!*

I refer to this as an Act of God. Out of respect for those who do not believe in God, may I suggest that you read this as Higher Power, HP.

No doubt, readers may well wonder just how people can make history, if they do not know what they are doing! Good question!

The best answer I can give is to draw your attention to the French Revolution of 1789. It was the first great revolution of the industrial age, and it was completely spontaneous. There were no leaders, although leaders did emerge. This in no way changes the fact that countless common people, both workers and peasants, rose up and overthrew the nobility!

The whole world was shocked! No one was worried about the nobility being overthrown, if only because no one even considered the possibility! The intellectuals, referred to as the "intelligentsia", were just as surprised as the nobility! Perhaps the people who were the most surprised, were the common people! The workers and peasants! They had no idea! Yet they rose up and made history! Proving once and for all, that truly, as Marx stated, "The masses are the makers of history"!

This brings us to our current situation. The working class is cultured, is in motion, and without doubt, a revolution is sure to break out soon. That *in no way* implies that the billionaires are sure to be overthrown, the existing state apparatus is to be smashed, and replaced with scientific socialism, in the form of the Dictatorship of the Proletariat. Such is hardly the case!

For one thing, a great many working class people are *not* convinced of the necessity of revolution. Further, they are certainly *not* convinced that it is necessary to *smash* the existing state apparatus, at the time of the revolution, and establish the Dictatorship of the Proletariat. Many still believe the lies of the billionaires, that we live under a democratic republic, and have "majority rule".

Of course, such is *not* the case, but we must respect their belief. They must learn, from their own *experience*, that the billionaires are in charge, and fully intend to remain in charge!

With that in mind, may I suggest that we "take the capitalists at their word". The capitalists say that if we do not like the system, we should "change it from within". Excellent idea!

In the interest of *raising the level of awareness of the working class*, of persuading so many working class people that the billionaires are in charge, I have a suggestion to make.

All Americans who would like to see change, from students to seniors, employed and unemployed, hungry and homeless, members of unions and sports clubs -*everyone!*- should join the two main stream political parties, Democratic and Republican, as *card carrying members!* This may be "against the rules", but there is no law against this! So feel free to be naughty! Now is not the time to be shy!

Those who consider themselves to be Independent Socialists, Social Democrats, Democratic Socialists or just plain Socialists, may find this to be most appealing. I can further suggest that you encourage all of your relatives and friends to also join the two parties. No doubt most of you are aware of the revolutionary theories of Marx and Lenin. You may or may not agree with them. Either way, is perfectly acceptable. We do not have to be in complete agreement, in order to work together.

Equally without doubt, all American Socialists are well aware that the Declaration of Independence guarantees all Americans, the right to "abolish any government which does not represent them"! Further, that same Declaration of Independence guarantees all Americans the right to "life, liberty and the pursuit of happiness". Go for it!

As card carrying members of the two parties, Democrat and Republican, you can suggest a new political platform, based upon the Declaration of Independence. Certain fundamental human rights should be guaranteed, such as medical, education, clothing, proper food, adequate housing and the right to live a life of dignity. How else can Americans have "life, liberty and the pursuit of happiness"? With that in mind, may I suggest the following:

1) Free medical for all, including emergency services, hospitalizations, home care and medications

2) Free education for all, no tuition, cancel all student loans

3) Free and nutritions food for all

4) Free housing for all, the abandoned buildings to be converted into housing

5) Cancellation of the national debt

6) Pensions for seniors to be tax free

7) Graduated income tax, with the rich paying their fair share

8) Military training for all able bodied people, both male and female, over the age of 16

9) People with such training to take over the role of the police and National Guard, on a part time basis, with the days worked to be paid for, by the employer

10) All elected officials to be subject to recall at any time, and paid the wages of workers

11) All banks in the country to be amalgamated, and put under workers control

12) All prisons to be shut down, the inmates to be put to work

No doubt, the vast majority of voters will embrace the above program. That includes those who are considered to be on the "Left", as well as those who are classified as "Right Wing", such as the supporters of Trump. After all, it is in their best interests.

Equally without doubt, all Socialists will think that this platform does not have a "hope in hell" of being carried out. Right you are! The party bosses, as the loyal and devoted servants of the billionaires, will not allow this! They will certainly *not* allow common people to become card carrying members of either party! Nor will they allow such people to set Party policy!

In this way, the less advanced workers will become convinced, from their own *experience,* that the Marxists are correct, that the billionaires are in charge, that there is no "majority rule".

On the other hand, it is entirely possible that, if enough working people place enough pressure on the Party bosses, that a few of them could conceivably become candidates for political office. If elected, they could even join the members of "The Squad", in Washington.

It is important to bear in mind that this is *not* a "numbers game"! We are *not* trying to flood Washington with socialists! We are *not* trying to take over the existing state apparatus! We *are* determined to *raise the level of awareness* of the working class!

Bear in mind that, under no circumstances, should any socialist run against Senator Sanders, or any member of The Squad!

Focus! The *goal* is to *raise the level of awareness* of the working class, the proletariat. If we succeed in sending a few people to Washington, fine. If not, that is also fine. The important thing is to give the working people *experience* in attempting to "change the system from within".

Those making this attempt will soon realize that the billionaires, working through their "party bosses", are in charge! There is *no majority rule!* They are not about to allow the working people to "take over" power in Washington!

In this manner, the working people will come to realize, from their own *experience,* that the party bosses are nothing but liars and swindlers! Devoted servants of the billionaires! "Majority rule" is a *myth!* The billionaires are *in charge!*

This will serve to drive the point home, to the working class, that the Communists are right. Democracy is *not* majority rule! Democracy is *class rule!*

This brings us to the subject of the workers who are somewhat more advanced. They may be well aware that the billionaires are in charge, and suspect that socialism is the answer, but are confused by all the political parties and groups, all claiming to be socialist. Such confusion is understandable!

For those working people, I am happy to say that the capitalists have thoughtfully provided us with the tools we need, in order to straighten out any confusion. The internet! That and digital devices, such as personal computers, smart phones and such, which most workers have in their possession. Or if they do not, then their children certainly do! And those kids can prove to be most useful! My grandchildren help me out constantly!

It is urgent that all workers read that most relevant book by Lenin, State and Revolution. He wrote that masterpiece in early 1917, in preparation for the Great October Socialist Revolution. He was determined that it would be a proper socialist revolution, and not a revolution which merely resulted in the transfer of power, from one set of rulers, to another.

That book was based upon the work which had previously been done by Marx, concerning revolution. In particular, Lenin paid strict attention to the article of Marx, titled, Civil War In France: The Paris Commune.

The workers and peasants of Russia took the advice of Lenin! For that reason, on October 25, old style calendar, or November 7, new style calendar, they mounted a successful insurrection! This has gone down in history as the Great October Socialist Revolution.

It is now up to us, to follow in their footsteps.

The approaching American Revolution is certain to resemble, quite closely, the Great October Socialist Revolution. For that reason, all people who are about to take part in that revolution, should read that book carefully, and take the advice of Lenin.

Revolution is not a tea party! Nor is it a game of cards, in which you lose one game, and win the next! It is a challenge to the authority of a *class* of people who are completely ruthless! The billionaires are determined to remain in power! To keep their wealth! If the insurrection is *not* successful, then the full wrath and fury of the entire state apparatus will be released, upon those who dared to challenge the authority of the billionaires!

The insurrection must be well planned and organized, nationwide. It must be carried out by people who are devoted to the cause of socialism! Such people must be prepared to *die for the revolution!* As I have covered this in a different article, there is no need to go into detail here.

Suffice it to say that *if* American revolutionaries follow the advice of Lenin, as detailed in State and Revolution, then the revolution will almost certainly succeed!

State and Revolution can be bought online, for a reasonable price. As well, it is now available in audio form, and can be downloaded from the internet. As well, Left Wing Communism, An Infantile Disorder, is also available in audio form. Another work of Lenin that I highly recommend! People can listen to these audios, perhaps while driving, performing house work, soaking up some sun, at work, or merely relaxing. I cannot imagine a better way to while away the time!

Now to return to our current situation, which is similar to the situation which existed in Russia, immediately before the Great October Socialist Revolution.

Since the time of Lenin, the level of awareness, of the working class, has regressed, largely due to the treachery of various leaders. Yet the sacrifices of so many proletarian leader, including the murders of Rosa Luxemburg and Karl Liebknecht, should inspire us to redouble our efforts. Let it never be said that they died for nothing! Let us walk proudly in their footsteps! Let us honour their memory, by overthrowing our current crop of billionaires!

Perhaps we can start by facing the fact that the combination of the Corona Virus pandemic and of the Second Great Depression -which the capitalists still deny!- has caused a crisis in capitalism, such as has not been seen since the First Great Depression, of the nineteen thirties.

We now have massive unemployment, homelessness, hunger, overdoses, suicides, mass shootings, drug and alcohol abuse, and open sales of drugs, on the streets! Criminal gangs are now terrorizing whole neighbourhoods!

The police are powerless, against such organized violence! Some cities are calling for the National Guard!

Yet the "mass movement", the revolutionary motion, is becoming ever stronger. Countless working people, those who were formerly apathetic, are now politically active. These are the advanced workers, the more enlightened members of the working class, the proletariat.

The American proletariat has now gone about as far as it can, on its own. Different groups are demanding different reforms. Black Lives Matter is demanding an end to violent police repression. Me Too is demanding an end to sexual assaults. The Autonomous Zone movement is demanding various reform. Across the country, people are becoming politically active, demanding change.

Now is the time for the working class to become aware of itself, *as a class*, with its own class interests. It must also become aware of the fact that it is necessary to overthrow the billionaires! This can only be accomplished through *revolution!* At that time, the existing state apparatus must be *smashed!* It must then be replaced with the Dictatorship of the Proletariat!

A careful reading of those two works of Lenin, State and Revolution, and Left Wing Communism, An Infantile Disorder, will serve to drive that point home!

It may help to think of the working class, the proletariat, as an army at war, which is precisely the case. The enemy is the capitalists, the billionaires. The enemy is also very strong, deeply entrenched. But then they have had many years to prepare their defences.

Yet, just because the "army" of the working class, far out numbers the "army" of the billionaires, does *not* mean that the working class has a huge advantage! Because the workers are *not aware* that they are even at war with the billionaires!

In previous writings, I have compared the class struggle to two boxers, one of whom is blind folded. The boxer who is blind folded, is of course striking out wildly, in all directions, and is sure to land a crushing blow, on occasion, strictly by chance. By contrast, the other boxer is able to land blow after blow.

As for those who may consider this to be a childish oversimplification -which it is!-may I suggest a comparison to that which is referred to as the "one six insurrection". As all Americans are well aware, on January 6, 2021, a disorganized mob, of possibly ten thousand working class people, stormed the capital building, in Washington. Clearly, those people were angry and frustrated, lashing out at their democratically elected politicians. A "blindfolded boxer"!

They were angry because as far as they were concerned, Trump had won the election! They were convinced that the election was fraudulent! It had been fixed!

From the response of the capitalists, as expressed in the mainstream press, it is clear that the working class, the "blindfolded boxer", had "landed a crushing blow"! Strictly by chance!

The working people, who were protesting, were right about one thing. The election was fraudulent! Joe Biden is a fraudulent President! Further, Kamela Harris is a fraudulent Vice President! The *federal election was fraudulent!* It was *Unconstitutional!* Countless members of the working class *sensed* this, with their *class instincts!*

As for those who are skeptical, may I refer you to the Twelfth Amendment, to the Constitution. It details the precise manner in which the President, and the Vice President, are to be elected.

In particular, the Twelfth Amendment states that:"The Electors shall meet in their respective states, and vote by ballot for President and Vice President...they shall name in their ballots the person voted for as President, and in distinct ballots, the person voted for as Vice President".

It does not take a Philadelphia lawyer to figure out, that there is *no* mention of any popular vote! The citizens have *no* voice in the federal election! There is *no* mention of any District! There is *no* mention of any political party! There is *no* mention of any "candidate"! There is *no* mention of any "running mate"! There *is* a mention of "distinct ballots"!

The Electors have every right to vote for the *"person"* of *their* choice for President, as well as the *"person"* of *their* choice, for Vice President! Those "persons" need *not* be the designated candidates, of any political party! Those "persons" need *not* even be a member of any political party! Further, the states have *no* right to meddle, in any federal election! The states have *no* right to *force* any Elector, to vote for *any* candidate, of *any* political party!

The 2020 federal election did not follow the procedure, clearly laid out in the Twelfth Amendment to the Constitution, so that federal election was *fraudulent!* It was *Unconstitutional!*

No doubt, the strongest supporters of the billionaires, will object that I am hardly an expert on Constitutional law. Guilty as charged! I am not even a lawyer! I welcome any challenge to any federal election, in a court of law! Allow the Supreme Court to decide!

Equally without doubt, those same billionaire boot lickers will point out, that the two party system has been around since the days of the Civil War! The current system, that of having each political party choose a candidate for President, is "time honoured". It is also a time honoured custom to have that particular candidate, of each party, to choose a "running mate", for the office of Vice President. It is further a "time honoured tradition" to have the candidates, of one or the other political party, elected to office! True! It is indeed "customary"! Even "time honoured"! That does not make it *legal!*

As for those who object that, "If it is Unconstitutional, then why has the Supreme Court not struck down those state laws, many years ago?" Because the Supreme Court rules *only* upon the Constitutionality of laws which are brought before it! It does *not go out of its way* to strike down laws, either state or federal!

Those state laws, requiring Electors to vote for particular candidates, chosen by a mainstream political party, has never been *challenged in court!* Those Electors who have chosen to *break* state laws, and vote for a different candidate, have *never been charged!*

The only reason the states have chosen *not* to charge Electors, who break state laws, is because state officials are *well aware* that those laws are *Unconstitutional!* They know full well, that if it ever gets to court, then those state laws will be *struck down!*

As far as the capitalists are concerned, this would serve only to "open a whole can of worms"! Reality check! That "can of worms" is nothing other than the *law!* The *Constitution!* The very Constitution that elected officials have taken an oath to *"preserve, protect and defend"!* Those same elected officials have a *duty to enforce those laws! To preserve, protect and defend the Constitution!*

If enforcing the law results in a ruling, by the Supreme Court, that the 2020 federal election was fraudulent, then *so be it! If* Joe Biden is a fraudulent President, and *if* Kamela Harris is a fraudulent Vice President, then *face it!* At the same time, in the case of such a Supreme Court ruling, feel free to face the fact that *all* federal elections, for perhaps the last *one hundred fifty years,* have resulted in *fraudulent Presidents,* and *fraudulent Vice Presidents!*

A great many working people, those who have faith in the democratic process, will find this to be completely shocking. To such people, I can only respond that such political shenanigans, such outright *fraud,* such *deception and deceit,* on the part of the billionaires, is not exceptional. It is *typical!* It is *bourgeois democracy!*

That is the reason I so highly recommend those two great works of Lenin, State and Revolution, and Left Wing Communism, An Infantile Disorder. A careful reading to those two books, will go a long way towards making all workers *class conscious! Aware* of the fact that the billionaires must be *overthrown! Aware* of the fact that this is possible only through *revolution! Aware* of the fact that the *existing state apparatus must be destroyed! Aware* of the fact that the existing state apparatus must be *replaced with the Dictatorship of the Proletariat!*

A careful reading of those two books will reveal, to the more advanced workers that, as Lenin pointed out, it was Marx who first proved that the class struggle, between the working class and the capitalist class, will of *necessity* lead to the Dictatorship of the Proletariat. This fact is accepted by all *scientific socialists!*

This is to stress the fact that Lenin was every bit the social scientist, as was Marx! Lenin was determined to *not* repeat the mistakes of previous revolutions, as was pointed out by Marx!

We can think of State and Revolution as a "road map" to revolution! Those who follow that "road map", in the forthcoming American Revolution, are sure to be successful!

CHAPTER 6

THE PROLETARIAN REVOLUTION AND THE RENEGADE KAUTKSY

enin also had his hands full with the social chauvinists. Immediately after the Russian Revolution of November 7, 1917, new style calendar, the most advanced strata of the proletariat, in many countries of the world, embraced Soviet (Council) Power and the Dictatorship of the Proletariat. In response to this, the social chauvinists launched a most vicious campaign of slander and distortion.

Perhaps the most skillful of these was a former Marxist, a *previously* highly respected individual by the name of Karl Kautsky. His pamphlet, The Dictatorship of the Proletariat, stands to this day as a masterful distortion of Marxism. As the American working class becomes familiar with Soviet (Council) Power and the Dictatorship of the Proletariat, we can expect the social chauvinists, to sing the praises of Kautsky. For that reason, it is best to give a little historical background.

In his early days, Kautsky was a fine theoretician, a superb Marxist. Yet that was at a time of relative peace, in the sense that the so called "Great

Powers", by which was meant the most highly industrialized countries of the world, were merely preparing to destroy each other.

These "Great Powers" were involved in a "rivalry in conquest", as by the turn of the twentieth century, the whole world had been divided up between them. The British were quite proud of the fact that "the sun never sets on the British Empire". They bragged of this constantly.

This merely had the effect of "rubbing salt into the wound", of the opposing "Great Powers". In particular, the Germans felt cheated. They had far fewer colonies than Britain, and thought it best to "level the playing field".

With that in mind, two huge world alliances took shape, one led by Britain, and the other led by Germany.

Britain and her friends became known as the Allied Powers, and included France, Italy, America, Russia, Romania, Japan and their colonies.

The opposing world power, led by Germany, included Austro-Hungary, Bulgaria, the Ottoman Empire, and their colonies. They became known as the Axis Powers. The Axis was determined to secure more colonies, while the Allies were determined to hang on to the colonies they had, and preferably add to them. It was a war to redivide the world, and has gone down in history as World War 1.

Caught in the middle were the common people, the workers and peasants. They were the "cannon fodder", and were sacrificed, by the millions.

In 1914, the tension reached a fever pitch, and the "Great Imperialist Slaughter" began. At the same time, the governments of each and every "Great Power", did the same thing! Each government called for the "Defence of the Fatherland"!

Most of the Marxists, those whom had previously been so zealously calling for revolution and the subsequent Dictatorship of the Proletariat, *collapsed* under pressure! They largely *converted to social chauvinism*, those who are socialists in words, chauvinists in deeds. That included Karl

Kautsky, a man who has since gone down in history as the "Benedict Arnold" of Marxism!

As mentioned earlier, his pamphlet, The Dictatorship of the Proletariat, was widely praised by the social chauvinists of the time. As very soon the American proletariat will also be discussing the forth coming revolution, Soviet Power and the Dictatorship of the Proletariat, no doubt the American social chauvinists will also be singing the praises of Kautsky. For that reason, it is perhaps best to discuss his pamphlet in more detail.

It was Lenin who responded to the "revisionist" efforts of Kautsky, with his own pamphlet, The Proletarian Revolution and the Renegade Kautsky.

As an aside, we should add that a "revisionist" is one who attempts to revise the revolutionary theories of Marx and Lenin.

Lenin documented the revisionist heresies of Kautsky in the following manner: "Kautsky's pamphlet, The Dictatorship of the Proletariat...is a most lucid example of that utter and ignominious bankruptcy of the Second International about which all honest socialists in all countries have been talking for a long time. The proletarian revolution is now becoming a practical issue in a number of countries, and an examination of Kautsky's renegade sophistries and his complete renunciation of Marxism is therefore essential."

The social chauvinists are quite predictable. They can be expected to rehash the same old garbage which has been spewed out many years ago. They will extol the book of Kautsky as a "model" of Marxist literature, as a "clarification" of the scientific theories of Marx.

Lenin had a few words to say, concerning the traitor to Marxism, the social chauvinist Kautsky.

In particular, Kautsky was enthusiastic about that which he referred to as "democracy as a method of majority rule"!

Lenin responded to this, in terms which left no room for any misunderstanding:

"In these circumstances, to assume that in a revolution which is at all profound and serious the issue is decided simply by the relation between the majority and minority is the acme of stupidity, the silliest prejudice of a common liberal, an attempt to *deceive the people* by concealing from them a well-established historical truth. This historical truth is that in every profound revolution, the *prolonged, stubborn and desperate* resistance of the exploiters, whom for a number of years retain important practical advantages over the exploited, is the *rule*. Never -except in the sentimental fantasies of the sentimental fool Kautsky- will the exploiters submit to the decision of the exploited majority, without trying to make use of their advantages in a last desperate battle, or series of battles." (italics by Lenin)

Lenin went on to state: "The transition from capitalism to communism takes an entire historical epoch. Until this epoch is over, the exploiters inevitably cherish the hope of restoration, and this *hope* turns into *attempts* at restoration. After their first serious defeat, the overthrown exploiters- whom had not expected their overthrow, never believed it possible, never conceded the thought of it- throw themselves with energy grown tenfold, with furious passion and hatred grown a hundred fold, into the battle for the recovery of the 'paradise', of which they were deprived, on behalf of their families, whom had been leading such a sweet and easy life, and whom now the 'common herd' is condemning to ruin and destitution". (italics by Lenin)

That is precisely the reason we need the Dictatorship of the Proletariat! Either we crush the billionaires, *after* the revolution, or they will return to power!

Lenin make this quite clear when he went on to state: "Kautsky... is a most typical and striking example of how a verbal recognition of Marxism has led in practice...into a bourgeois liberal theory recognizing the non-revolutionary 'class' struggle of the proletariat...By means of patent sophistry, Marxism is stripped of its revolutionary spirit; *everything* is recognized in Marxism *except* the revolutionary methods of struggle, the propaganda and preparation of those methods, and the education of the masses in this direction. Kautsky 'reconciles' in an unprincipled way the fundamental idea of social chauvinism...The working class cannot play its world revolutionary role unless it wages a ruthless struggle against this backsliding, spinelessness, subservience to opportunism, and unparalleled vulgarization of the theories of Marxism". (italics by Lenin)

I have chosen to go into this in detail, as it is so vitally important. After all, the capitalists managed to return to power, in both the Soviet Union and China. In those formerly socialist countries, at the time the working class was in power, the proletariat did *not* exercise *sufficient* dictatorship over the capitalists.

As is well known, at the time of Lenin, a number of countries, especially in western Europe, were close to revolution. Yet that revolution never took place, if only because the working people were deprived of their leaders. Most of the Marxists took the lead of Kautsky and became social chauvinists, devoted servants of the capitalists.

There were notable exceptions, such as Rosa Luxemburg and Karl Liebnecht of Germany, fine Marxists, people of principle. As such, the capitalists recognized them as the threat they were. The capitalists murdered them.

It is true that in various countries of western Europe, during the time of Lenin, numerous uprisings took place. These were isolated and scattered incidents, lacked focus and direction, as there were no Marxist leaders to provide the proper direction. This merely confirms the statement of Engels, to the effect that "without a proper revolutionary theory, there can be no revolutionary motion".

Modern day Marxists, otherwise known as Communists, would do well to bear this in mind! While putting forth the revolutionary theories of Marx and Lenin, they would be well advised to take reasonable precautions!

We now have some fine tools, as modern technology has provided us with various electronic digital devices, as well as the internet. We can use these tools to communicate with each other, as well as with countless working people. At the same time, we do not want the government officials knowing who we are.

The "perverts", including child molesters and human traffickers, manage to stay in touch, using something referred to as the "dark net", so clearly, it can be done.

We can expect the capitalists to perpetuate the myth of "pure democracy", as opposed to a dictatorship, and in particular the Dictatorship of the Proletariat. It is the duty of true Marxists, Communists, to make the working class aware of the fact that "pure democracy" is an oxymoron, a contradiction in terms. Democracy is merely a state apparatus, a method by which one class suppresses, crushes, another class. There is nothing "pure" about a class being crushed! Under bourgeois democracy, the tiny class of the minority, the billionaires, rules. They in turn crush the vast majority, the working class, the proletariat. They also exploit the proletariat. It is democracy for the billionaires, but a dictatorship for the proletariat.

By contrast, under the Dictatorship of the Proletariat, the working class, the proletariat rules. It is democracy *of* the working class, *by* the working class, *for* the working class! For the capitalists, the billionaires, it is a dictatorship! That is a fact!

It is also a fact that, after the revolution, the working class will establish working class courts, complete with working class judges. It is very likely that three workers will serve on each court. Elected and subject to recall at any time. Their decisions will be based on *common sense*, not on law books. Those who are convicted, can expect to be sentenced to terms of manual labour. They will not be going to a country club of a prison!

The criminal elements, the thieves and killers, the dope dealers, the mobsters, and in fact all the criminal gangs that now terrorize the country, will have *no rights*! They too will be *crushed!* They will *not* be allowed to hide behind high priced lawyers! They will *not* be released on technicalities! Once convicted, they will *not* be provided with free lodging, medical, dental, vision and hearing! They will *not* receive free recreation, education and entertainment! They will *not* be sent to a country club of a prison! They will *not* be molly coddled! They *will* be put to work! They *will* be forced to perform honest labour! For possibly the first time in their lives! No more statute of limitations! They *will* experience the full weight and fury of the Dictatorship of the Proletariat!

When appropriate, before serving their sentences, they will first be paraded through the streets of the neighbourhoods they have perviously terrorized. The people who lived in such mortal dread, all their lives, of these mobsters, will be allowed to pelt these scum with garbage! In this way, the *spiritual power* of the mobsters will be broken.

As for the sex offenders, they will be forced to face the women they have assaulted! The "human traffickers" will face the females that they have bought and sold! The "pimps" and "madams" will face the women they have been selling! The pedophiles will face the mothers of the children they have molested!

In each case, the women, the *victims* of these human scum, will decide the punishment of their abusers! In this way also, the *spiritual power* of the sex offenders, will be broken! Those women will finally secure the justice, they have so long been denied!

Only as a last resort will the death penalty be applied. After all, those who are dead, can no longer be the slightest bit useful.

No doubt, the social chauvinists will complain that, "This is not right! Innocent until proven guilty"! To this we can only respond that under the Dictatorship of the Proletariat, it is the *victims* who have all the rights! For criminals and thieves, it is a Dictatorship*!*

This stands in stark contrast to our current situation, under capitalism, in which the criminals have *all* the rights! Victims have *no* rights!

As for those who are skeptical, consider that which is happening now.

Criminal gangs are currently running wild, terrorizing whole neighbourhoods. They have a certain *spiritual power* over the working people. Common people are terrified of them, with good reason! The mobsters refer to this as "respect". This spiritual power must be *broken!* This can be accomplished by *humiliating* those mobsters, parading them through the streets, and allowing the working people to throw garbage at them!

It is only *after* the mobsters are humiliated, *after* their spiritual power is broken, that they will be forced to serve their sentence, possibly a lengthy period of manual labour. Or possibly climbing the thirteen steps of the gallows!

No doubt, the bleeding heart liberals will shed "crocodile tears" for those "poor unfortunates", those thieves and killers, those rapists and pedophiles! Denied their "democratic rights"! Typical Kautskyites!

The fact of the matter is that the working class, the proletariat, *must liberate itself!* That is a fundamental tenet of Marxism! It is also a fact that spiritual power is every bit as real, as the physical force elements of repression! Just because it cannot be seen, does not mean that it does not exist! Further, just as the physical force elements of repression must be *destroyed*, so too, the spiritual force element must also be *destroyed!* The working class must take part in their own emancipation! This can only be accomplished through the humiliation of those who have terrorized them!

At the time this is happening, we can expect the Kautskyites to complain that this humiliation is "terrible"! We can also expect the more advanced to say that "it may be terrible, but it is necessary"! This is closer to the truth, but they too are mistaken. The humiliation and degradation of those who have crushed and exploited the working class, is necessary! It is wonderful! It is to be embraced! In no other way can the working people "break the invisible chains" of spiritual oppression!

During a time of revolution, it is sometimes necessary to exceed the bounds of decency, to step outside the bounds of that which is usually considered to be socially acceptable behaviour. In no other way can the working class liberate itself.

CHAPTER 7

INSURRECTION IMMINENT

ithin America, the tension is mounting, daily rising to greater levels. Desperation is wide spread. The gun violence is reported to be at "epidemic proportions", with senseless shootings, "mass casualty events", almost on a daily basis. Yet history reveals that Americans are not about to tolerate this. This has been demonstrated in the past. Americans can be pushed only so far!

Around the year 1863, in that which is commonly referred to as the "old west", there was a well-organized gang of thieves which was stealing from a group of miners, hardworking people, who would dig the ore out of the ground, only to have it stolen. All of these miners were honest men, some of whom worked their own claims, while others worked for wages. They all did what they had to do, in order to survive.

The local law enforcement was stumped, the sheriff unable to track down these outlaws. He was dead set opposed to the creation of any "vigilante committee", as it was up to the *law* to apprehend these bandits, and bring them to justice. Citizens do *not* have the right to take the law into their own hands!

Yet the honest, hardworking citizens decided that "enough was enough", and formed a vigilante committee. They then grabbed a man who always seemed to have money, even though he never worked. They suspected that he was one of the gang, and they were right. The citizens were able to extract information from this man, concerning the identity of the other gang members.

The leader of the gang was a man named Plummer, who just happened to be the Sheriff of the town! No wonder he could not catch these outlaws! He was the head of the gang! So the citizens, the vigilantes, grabbed Plummer and perhaps thirty of the gang members, or at least all they could get their hands on, and hung them. Frontier justice! They were strung up from the nearest tree.

Other gang members high tailed it out of there, quite horrified that there was no trial, no judge and jury, no lawyers to hide behind. This was an early demonstration of American justice, treating thieves and killers in the proper manner.

Now we have a similar situation in America. Once again the thieves and killers are running amok, terrorizing and robbing the working people. Only this time the gang is a class of people, the billionaires. Once again we can expect the workers to come together, only this time it will not be as local vigilantes, but as a nationwide network, the working class united in their opposition to those thieves. This network of workers must plan an insurrection, in order to overthrow the billionaires and establish a socialist republic, in the form of the Dictatorship of the Proletariat.

With that in mind, perhaps a little word from Lenin: "To be successful, insurrection must rely not upon conspiracy and not upon a party, but upon the advanced class. This is the first point. Insurrection must rely upon a *revolutionary upsurge of the people.* That is the second point. Insurrection must rely upon that *turning point* in the history of the growing revolution when the activity of the advanced ranks of the people is at its height, and when the *vacillations* in the ranks of the enemy and *in the ranks of the weak, half-hearted and irresolute friends of the revolution are strongest.* That is the third point. ...Once these conditions exist, however, to refuse to treat insurrection as an *art* is a betrayal of Marxism and a betrayal of the revolution." (italics by Lenin)

The current situation is revolutionary. The middle class, the petty bourgeois, is now being "wiped out," by the billionaires. Those still in business are "living on borrowed time".

The billionaires are *ambitious!* The billionaires are *focused!* The billionaires are *determined1* The billionaires are *not* content to remain billionaires! The billionaires are *focused* and *determined* to become *Trillionaires!*

It is simply a matter of wiping out the middle class, some of whom have "assets" worth billions, of seizing those assets, and presto! Instant Trillionaires!

We know this for a fact, because the billionaires were kind enough to tell us! They recently announced that *certain* businesses are *Too Big To Fail!* They gave a list of only *eight* American banks that meet that requirement! One of those eight banks is Wells Fargo. It has the "least" amount of assets, worth a "mere" *1.88 Trillion* dollars!

It stands to reason that *a Trillion* is the dividing line! Businesses with assets of *less* than a Trillion, are *Too Small To Succeed!* As those "small businesses" collapse, those that are "Too Big To Fail", merely grab their assets!

As for those who are skeptical, who may think that perhaps the billionaires may be joking, feel free to consider the fact that First Republic Bank, with assets of *212.6 billion,* recently failed. Even a "cash infusion"of *one hundred billion,* failed to save that bank! After all, it was clearly "Too Small To Succeed"!

Those assets were promptly "picked up", by a bank that tops the list of those that are Too Big To Fail, with assets of *3.74 Trillion!* That bank is JP Morgan Chase.

This is to say that the members of the class of people known as the "upper middle class", no longer have a reason to feel "safe and secure"! The billionaires have just announced that it is now "open season" on them!

To the upper middle class people, those who formerly had "visions" of joining the billionaires -who can blame them?- may I suggest that you cast aside those illusions! That is not about to happen! Your days are numbered! You are living on borrowed time! *Cut your losses!*

Three banks have recently collapsed! The "tip of the iceberg"! *Several thousands* of others are about to join them! They will in turn drag down countless "small businesses", those that are Too Small To Succeed! *Depression!* The *Second Great Depression!*

Everyone is no doubt well aware of the assurances, of the servants of the billionaires, to the effect that "the first quarter millions of all deposits are insured by the FDIC". That used to be true! Now, it is almost certainly a lie!

The fact is that the "Fed" just gave *one hundred ten billion* to two banks that were failing, Silicon Valley and First Republic. Pouring water down a dry well! Guess where the head of the Fed got that money! Feel free to ask her!

Even now, there are certain middle class people who are "liquidating", selling whatever they can, raising "cash money", so that they can buy gold and silver, and place these "solid assets" in a "secure location", probably safe deposit boxes.

That in no way *solves* the problem of monopoly capitalism! The *billionaires* are the problem! They must be *overthrown!* Capitalism must be replaced with *socialism!* That calls for a *revolution!* The only *class* of people who are capable of mounting an insurrection, overthrowing the billionaires, and establishing a socialist society, is the *proletariat!*

Allow me to stress the fact that it is in the *best interests* of middle class people, to work towards socialism! You have no future under capitalism!

The billionaires are determined to drive you into *bankruptcy! Financial Ruin!* Yet you have a bright future under socialism!

Under scientific socialism, the Dictatorship of the Proletariat, professional people will be in demand! That includes business people! As that is the case, they will be paid most handsomely!

The infrastructure of the country is in a state of decay! It is falling apart! The roads, bridges, tunnels, railroads and a great many buildings, have to be repaired or replaced. Numerous railroads have to be built, repaired and rebuilt.

Countless businesses, which have gone bankrupt, have to be re-opened. The areas which have been severely polluted, have to be cleaned up. This can only be accomplished with the expertise of business managers. After all, there is a big difference between running a machine, and running a business!

With that in mind, may I suggest that such people take to heart that ancient expression, "The Lord helps those who help themselves!"

There is no need to wait for the banks to fail, for the stock market to collapse, for all small businesses to "crash and burn"! Take the "bull by the horns"! Get active! Get involved with the Councils (Soviets) which are springing up, all across the country! Take part in arming, equipping and training the working people, in preparation for the approaching revolution. The insurrection is critical! The events of "one-six" are a fine example of how *not* to mount an insurrection!

As well, the working people, or at least the most advanced strata of the working class, are now raising their level of awareness, by reading those two critically important works of Lenin, State and Revolution, and Left Wing Communism, An Infantile Disorder.

That is excellent, but fails to face the fact that all people need leaders. The working class is no exception. The *only* political party which can provide that leadership, is a true Communist Party, one which calls for the Dictatorship of the Proletariat.

Yet it is not reasonable to expect working people to create a true Communist Party. It is very likely beyond their ability, although no doubt certain working class intellectuals can be helpful.

The creation of such a Communist Party is clearly *not* beyond the ability of middle class intellectuals! On the other hand, do *not* make the mistake that Lenin made! Immediately after he created the Saint Petersburg League of Struggle for the Emancipation of the Working Class, he and all the other members, were promptly thrown in jail!

Be discrete! Use the internet! There is no need to meet in person! Avoid the telephone! The various intelligence networks monitor all phone calls, "flagging" certain words! Yet if the sex offenders can keep in touch, while escaping detection, then it can be done!

There is a sense of urgency to this. The mass movement of the working class is high. Desperation is gripping the country. The ruling class, the billionaires, are deadlocked. Washington is in a state of gridlock. The Democrats and Republicans are at each other's throats. About the only thing they can agree upon now, is to stop Trump. But they cannot even agree on *how* to do that!

Trump and a great many of his followers are convinced that he won the 2020 presidential election, and are determined to place Trump back in the White House, and soon. His plan is to be in the White House after the next 2024 federal election.

That could happen! Even though he is currently facing numerous criminal charges, with no less than *four* trials pending, Trump still has a most impressive following! Trump is a *leader!* Trump has *charisma!* Trump has *charm!* Trump is able to inspire *devotion* in others!

Trump is following in the footsteps of such American legends as Billy the Kid, Jesse James, Cole Younger, Wes Hardin, Scarface Al Capone, Machine Gun Kelly, Baby Face Nelson, Pretty Boy Floyd, Dutch Schultz, John Dillinger, Sammy the Bull and John Gotti. Leaders! People followed them! To prison! To the gallows! To the grave!

People need leaders! *Proper leaders! Not* thieves and killers! *Not* sex offenders! *Not* men who brag about grabbing women by the (genitals)! *Not Trump!*

Yet that is precisely what we have! And what are working people being offered, as an alternative? A doddering old man of a President, battling dementia! A Speaker of the House, who is a wimp! A Vice President, who is doing a most impressive Sarah Palin imitation!

With that in mind, may I suggest that there are numerous people who have the *potential,* to become *proper* leaders. This is *not* to say that we need people with "charm and charisma" to become leaders! Such qualities are definitely an asset, but the *important* thing, to become a *proper* leader, is the ability to *raise the level of awareness of the Proletariat,* to the level of *class consciousness!* Working class people must become *aware* of the revolutionary theories of Marx and Lenin! They must understand and *embrace* Soviet Power and the Dictatorship of the Proletariat! Those who are able to bring, to the common people, that *awareness,* are true *leaders!*

Most of those who are potential leaders, are "holding back", perhaps afraid to "stick their necks out". Intimidated, to phrase it politely. Not at all anxious to become martyrs!

To such people, I can only say that there are times, when it is necessary to *take a stand!*

Feel free to take inspiration from such people as the 300 Spartans, who made their stand at Thermopylae. They faced the whole Persian invading army! They held that important pass, between the mountains, for two days! A critical "rearguard" action! This gave the bulk of the Greek army time to retreat and regroup! Those Spartans died, to a man!

This is not to say that the Spartans were pleasant people. They were not! Yet they did their *duty!* No one can take that away from them!

Take a stand! Let your voice be heard! Challenge the billionaires! They must be stopped! Only the revolutionary working class, the proletariat, can stop them! Only through revolution! Only through the subsequent Dictatorship of the Proletariat! Only with the proper leaders!

As it stands now, it is very likely that Trump will win the Republican nomination for President, and run for office, in 2024. Or he may run as an Independent! His lawyers are working on postponing all trials, until after the federal election. They will probably succeed, as the law applies only to working people. The billionaires can do as they please!

Assuming that is the case, if he runs again for president, he will likely win! After all, given a choice between Trump and Biden, not too many people are going to vote for the one who has dementia!

This may well have the effect of triggering a proper revolution! The only question is, what then?

If the insurrection is well planned and executed, then the seizure of power can be expected to be almost bloodless. That was precisely the case of the Great October Socialist Revolution, but only because the revolutionaries took the advice of Lenin.

We can also expect it to be *led by women*, as the American working men have *not* fulfilled their *duty! Shameful!* By contrast, the women have proven themselves to be excellent organizers. We can only hope that the men will follow their lead. On that day, the capitalists will learn the meaning of the word *insurrection*! It involves something more than breaking a window and knocking down a cop!

On the day of the insurrection, the first day of the revolution, we can expect the flunkies of the capitalists, as well as the police, to have a ''conversion'', so that they will instantly be converted into die hard revolutionaries! They will abandon the capitalists in droves! They will approach us with outstretched hands, warm smiles on their faces! We in turn must embrace them as the Brothers and Sisters, the Comrades that they are! Their past service to the capitalists must not be held against them! We can use all the help we can get! Besides, the capitalists have trained them well! They have certain skills which, no doubt, will prove to be quite useful!

By contrast, on that day the billionaires will be completely stunned, alone and isolated, for perhaps the first time in their lives. They will cry out in vain for their servants. Slowly, the horrible, sickening realization will dawn on them, the fact that the unimaginable has happened. The unthinkable. The "plebians" have rebelled! The "lower classes", the "rag tag and bob tail", the "scum of the earth", the "dregs of humanity", the "gutter sweepings", the "trailer court trash", have revolted! Their whole world, that of luxury and decadence, will come crashing down!

Most will resist the first mad impulse to end it all, to commit suicide, although a few will succumb to temptation, just as so many Nazis ended their own lives, as the Third Reich came to an end. The more stalwart will brace themselves for the coming battle, against the Dictatorship of the Proletariat. They will be determined to restore their "paradise lost", to return to their life of luxury. It is simply a matter of overthrowing the Dictatorship of the Proletariat!

We can count on the billionaires to make every effort to return to power! This the capitalists managed, in the Soviet Union and in China. The fact is that both Stalin and Mao, although great revolutionaries, made mistakes. Those mistakes allowed the capitalists to return to power.

We must *learn* from the experience of Stalin and Mao, both their successes, and their mistakes! This is referred to as the proper application of scientific socialism! In this way, we will learn to exercise complete dictatorship over the billionaires! No half measures! Total Dictatorship of the Proletariat!

I have chosen to document this, in considerable detail, as it is so important. The point is that the only true Marxists, scientific socialists, are those who call for the Dictatorship of the Proletariat.

Now to return to the time of the approaching American Revolution.

After the insurrection, a new government will be established. The Councils, or Soviets, which have been spontaneously created, will form the new government, in the form of the Dictatorship of the Proletariat.

Incidentally, it remains to be seen if these newly created Councils will continue to be referred to as Councils, or as Soviets. That is entirely up to the working people.

The new government will, of necessity, be a coalition government, one which represents different classes, the proletariat and the petty bourgeois, as well as different factions within those classes. After all, not everyone embraces the Dictatorship of the Proletariat!

As the Russian Great October Socialist Revolution of 1917 was similar, perhaps it would be helpful to make a comparison.

In Russia of 1917, the situation was far more complicated. There were more political Parties, representing more classes, with factions within each Party.

The Social Democrats, the Marxist Party, was divided between the Bolsheviks, the Party of Lenin, and the Mensheviks, those who were revisionists, of the opinion that the theories of Marx should be revised.

The Constitutional Democrats, the Cadets, supported the nobility and the landlords, as the landlords were related, or at least devoted, to the nobility. They also generally supported the capitalists, the bourgeoisie.

The Socialist Revolutionaries tended to represent a great many peasants, although they were generally divided between the Left Socialist Revolutionaries, who were close to the Bolsheviks, and the Right Socialist Revolutionaries, who were close to the Cadets.

Even within the Bolshevik Party, there was Trotsky and his followers, who did their best to sabotage the revolution.

It did not help that Russia was also at war with Germany and the Central Powers, as well as being surrounded by hostile forces. Any gambler would have offered odds, heavily against a successful revolution. Yet the revolution succeeded, as Lenin was able to persuade so many people, of the correctness of the revolutionary theories of Marx. Perhaps those taking part in the current American revolution, can take inspiration from this.

Immediately after the November 7 insurrection, the Cadets demanded that the secret treaties of the Czar, with the British and French, be honoured. In this way, they exposed themselves as the strongest supporters of the capitalists. They wanted no part of the new socialist government, which was at that time being created.

In addition, the Right Socialist Revolutionaries and Mensheviks wanted to recognize the authority of the Kerensky Provisional Government. When this was denied, they walked out of the meeting.

As well, the Bolsheviks endorsed the agrarian program of the Socialist Revolutionaries. Not that they agreed entirely with it, but had no doubt that the peasants would learn, from experience, that the Bolsheviks were correct. This is referred to as compromise, but not on principle. At no point did the Bolsheviks compromise their principles.

At the same time, the Russian Empire collapsed. Numerous republics, which had been crushed under the Russian Empire, declared their independence. Many of them became separate independent socialist republics. Several years later, they joined Soviet Socialist Russia, to form the Union of Soviet Socialist Republics.

Here in America, at the time of the approaching revolution, we can expect the American Empire to also collapse. Even now, no less than three independent republics have taken shape. On the east coast, seven states have come together. In the midwest, the industrial heartland of the country, seven other states have also come together. And on the west coast, three states have merged.

Each of these areas will almost certainly declare themselves to be independent, separate republics. They will also very likely declare themselves to be socialist, with Councils in charge, the Dictatorship of the Proletariat.

As well, we can expect the colonies of Alaska and Hawaii to declare independence. In response to those who say that Alaska and Hawaii are states, I can only respond that they are *states in name only!*

That brings us to areas which are referred to as "Districts" or "Protectorates". The Districts include the District of Columbia, Puerto Rico, Guam, Northern Mariana Islands, American Samoa and US Virgin Islands. The "Protectorates" include Baker Island, Howland Island, Jarvis Island, Johnston Atoll, Kingman Reef, Wake Island, Midway Islands, Navassa Island and Serraiilla Bank.

These are all colonies, in everything but name! At the time of the next American Revolution, they will all achieve independence.

We can also expect a popular political party to take shape, in the form of a Social Democratic Party. Those who refer to themselves as independent socialists, or social democrats, can be expected to form this Party. Among the leaders we can expect to find Senator Sanders from Vermont, as well as those who are referred to as members of "The Squad", within the House of Representatives. They are well respected and will no doubt serve with distinction, within the new socialist government.

The true Marxists, Communists, those who call for the Dictatorship of the Proletariat -as that is the touchstone of a true Communist!- will no doubt work closely with these utopian socialists. We certainly have no quarrel with them! We respect their beliefs! They are people of principle!

The same cannot be said of the social chauvinists. They claim to be Marxists, while denying the necessity of the Dictatorship of the Proletariat. Completely devoid of principle! Opportunists! We want *nothing* to do with such people!

Immediately after the American insurrection, we can expect the strongest supporters of the capitalists, to demand that the national debt be honoured! If they cannot prevent the breakup of the American empire -and they cannot!- then the least they can do is demand that the separate socialist republics pay "their share" of the national debt! Not likely!

We can also expect those same boot lickers, of the billionaires, to demand that a "democratic republic" be established, along the same lines of the previous republic, with a President, Cabinet, Senate and Congress. Of course the plan of such people, is to set themselves up as the new rulers! Out of the question!

After some time, perhaps several years, we can expect several of those newly created socialist republics to come together, to form a modern Soviet (Council), Union. They could even join with other socialist republics, in other parts of the world. A World Socialist Republic.

For the moment, we have to face the fact that we do not have a proper Communist Party, Dictatorship of the Proletariat, or CP,DP. But as Lenin said, a successful insurrection does not rely upon a Communist Party. Nor does it rely upon a conspiracy. It relies upon the working class, the proletariat, the one and only advanced revolutionary class.

Still, even though a proper Communist Party is not absolutely necessary, the fact remains that working people need proper leaders. It is difficult to imagine a revolution being successful, overthrowing the billionaires, smashing the existing state apparatus, and setting up the Dictatorship of the Proletariat, without a proper Communist Party! Yet any other outcome to the revolution, would merely result in more capitalism, with a different group of rulers!

Bear in mind, that Lenin was also of the opinion that a very strong revolutionary movement would almost certainly give rise to a true Communist Party. Let us so hope!

In preparation for the revolution, the creation of "Autonomous Zones" is to be encouraged, but *not* with a view to declaring independence! The experience of the Seattle Autonomous Zone has proven that the capitalists will *not* tolerate any Zones which claim to be Autonomous! They see such Zones as a threat to their rule, which is precisely the case!

Such Zones must work "underground". The leaders must ensure that the members are trained, armed and equipped. Such equipment includes helmets, shields, bullet proof vests, night sticks, gas masks, paint ball guns and sling shots, complete with marbles.

As well, each group must be provided with pipe wrenches. The water trucks, complete with water cannons, get their water from the fire hydrants. The pipe wrenches must be used to open all fire hydrants in the city. With all the hydrants open, the water pressure drops to zero, and the water trucks are unable to secure any water. In that way, the water cannons will be neutralized.

It is also a fact that, with a little practice, a person can become surprisingly accurate with a sling shot, using marbles. These can be most effective against dogs and horses. Not that those marbles can be expected to kill horses, but a horse can be expected to give the rider flying lessons, when pelted with marbles. Even simple spears, three meter or ten foot wooden shafts, complete with metal tips, can prove to be most effective. After all, it has been known, since the time of Alexander the Great, that horses will not charge into a wall of spears. As well, everyone should receive military training, including the use of firearms.

Bear in mind that if Trump manages to get back into power, then the revolution will be much more difficult. The working class has to organize an insurrection, and soon. The only people who have proven themselves capable of this, are the women. The Women's March of 2017 was a spectacular success. Many millions of people took part that day, mainly women, all across the country.

Congratulations ladies, and now it is time to repeat the performance, but on a larger scale. Now it is time to mount a nationwide insurrection. The billionaires have to be overthrown, a socialist republic has to be established, and "fate" has decided that it is up to you ladies to take the bull by the horns. By fate, I mean that the working class men have failed to rise to the occasion.

I say this as a dedicated male chauvinist, only because it is true. Shameful!

For the moment, may the posters and banners read:

Scientific Socialism!

Victory or Death!

Fight Like A Girl!

Dictatorship Of the Proletariat!

Workers of the World, Unite!

CHAPTER 8

COLLAPSE OF THE AMERICAN EMPIRE AND NATIONAL LIBERATION

he revolutionary motion in America continues to intensify. The journalists are reporting that, since the beginning of the year, there have been over two hundred fifty mass shootings in the country. There were ten mass shootings last weekend alone. As well, a federal court has just struck down a California law, banning assault rifles. The sense of desperation within the country is being transformed, into a sense of despair! Even the most distinguished politicians are expressing fear that their democracy is about to collapse.

Their fears are well grounded! Their democracy *is* about to collapse, the democracy of the *billionaires!* It is *bourgeois* democracy, a democracy *of* the billionaires, *by* the billionaires, *for* the billionaires. It is about to be replaced by a *new* democracy, a *proletariat* democracy, a democracy *of* the proletariat, *by* the proletariat, *for* the proletariat. For the billionaires, it will be a *Dictatorship!* The *Dictatorship of the Proletariat!*

The country can best be thought of as a "powder key", in that any spark can cause an explosion. Such a spark is bound to give rise to a *spontaneous* uprising. What we need is a *conscious* uprising!

The problem is that the working class is *not yet* aware of itself *as a class!* With its own *class interests!* It is certainly *not* aware of the fact that it is *destined* to overthrow the completely reactionary class of monopoly capitalists, the billionaires! That is about to change, as countless working people, or at least the more advanced workers, are about to read two key works of Lenin, State and Revolution, and Left Wing Communism, An Infantile Disorder.

As well, now that so many middle class people have been ruined, forced into the ranks of the proletariat, they will no doubt perform a valuable service, for the working class. They too, will help to raise the level of awareness of the working class. After all, it is only members of the middle class who have been exposed to the revolutionary theories of Marx and Lenin, in university.

Such people also have the personal *experience,* of working with, or for, the billionaires. They can testify to the fact that Marx and Lenin were *correct!* They can further testify to the fact, that the billionaires are *well aware* that Marx and Lenin were correct! Which changes nothing! The billionaires are still determined to live their lives of luxury, while paying no taxes! Even though they are well aware that they are driving the country to *ruin!* There is a big difference between *knowing and caring!* And the billionaires simply *do not care!*

Incidentally, it is a mistake to think of Trump as a "bad" billionaire. Nor is he a "good" billionaire. He is merely a *typical* billionaire! He cares only about himself, and his immediate family! *No one else! All* billionaires feel the same way!

The combination of the Second Great Depression and the Corona Virus, has served to *accelerate* the course of world history. It has given rise to crises, such as have not been seen since the nineteen thirties. In America, the number of unemployed is at a level last seen during the Great Depression. Countless people are homeless, sleeping on the streets and under bridges. The "more fortunate" are able to live in vehicles. Millions of others are waiting for the temporary ban on evictions to expire, after which they too, will be evicted. The added burden of homelessness will then be added to their miserable lives.

This "crisis in capitalism" is causing a great many working people to fall into a deep depression. They are "self-medicating" with alcohol and drugs. Others are resorting to "random acts of violence", or "wilding", as is the popular expression. Still others are striking out with firearms, killing everyone they can! Overdoses and suicides are now common place! Criminal gangs rule entire neighbourhoods! Drugs are being sold openly! The police are powerless against heavily armed gangsters! Certain cities are calling for the National Guard!

That is "one side of the coin", so to speak. In scientific jargon, we say that this is "one aspect of the contradiction". But as "every coin has two sides", let us examine the "other side of the coin", or the "other aspect of the contradiction".

A great many people have "risen to the occasion"! They are determined to *stop* the billionaires! The billionaires will *not* be allowed to run this country *into the ground!* They are taking *action!* Certain things *must* be done! These actions may not be entirely *legal!* This does not change the fact that, legal or not, *certain things just have to be done!*

Under capitalism, it is sometimes *necessary and correct to break certain laws!* Not for personal gain, of course. And certainly not in the interests of spreading terror!

Across the country, numerous Councils (Soviets) are taking shape. (It remains to be seen which name will become commonplace.) The members of these Councils are deeply concerned with working people. For that reason, they are providing the desperately poor with food, clothing and necessities of life, such as soap, towels and blankets. Frequently, the items they are "stealing", including food, is that which is meant to be destroyed! The capitalists would rather destroy food, than give it to the hungry!

As well, certain Councils are providing shelter for battered women. These abused housewives are being re-located to secret locations, so that their husbands cannot find them. These Councils are taking care of such women, if only because the social agencies either cannot, or will not!

These are just a few examples of Councils performing valuable services, which is *against the law,* under capitalism! This is in addition to arming, equipping and training a great many people, in preparation for the approaching revolution!

Under these circumstances, with this crisis in capitalism, it is sometimes difficult to determine the proper thing to do. There is so much that *should be done,* and only so much that *can be done!*

For that reason, I have found the advice of Lenin to be most helpful. As he stated in What Is To Be Done?, "Political life as a whole is an endless chain consisting of an infinite number of links. The whole art of politics lies in finding and taking as firm a grip as we can of the link that is least likely to be struck from our hands, the one that is most important at the given moment, the one that most of all guarantees its possessor the possession of the whole chain".

In my opinion, the "key link" now, in the political chain, the link that is "most important", is that of *raising the level of awareness of the working class!* The working class must be *prepared for Soviet Power and the Dictatorship of the Proletariat!* The most advanced workers must become aware of the *revolutionary theories* of Marx and Lenin! They must be raised to the level of *true Marxists! Communists!*

In *no other way* can the billionaires be overthrown, and a scientific socialist society be established! Otherwise, the next revolution could succeed only in transferring power, from one set of rulers, to another!

It is clear that the current situation bears a striking resemblance to Russia, in 1917. In fact, almost all scholars are of the opinion that the approaching American Revolution will closely resemble the Great October Socialist Revolution.

For that reason, perhaps it is best to examine the Great October Socialist Revolution, as well as the events which led up to that Revolution. In this way, we can know what to expect. We can also learn to avoid the mistakes of those Revolutionaries.

In 1917, there was a considerable amount of industry in Russia, so that there was a rather sizeable working class, proletariat. This is of the utmost importance, as it is the proletariat which is the only *consistently revolutionary class!* The other classes, including the peasantry and middle class, tend to vacillate, between the capitalists and the proletariat.

By 1917, Russia was a three year veteran, and an enthusiastic participant, in the Great Imperialist Slaughter, that which has gone down in history as the First World War. The suffering of the common people, the workers and peasants, reached horrendous proportions.

The troops at the front were ill equipped, hungry, dressed in rags and frequently barefoot. Their morale was at rock bottom.

The people in the cities and countryside, were faring little better. Everyone wanted peace. Everyone was hungry. Everyone wanted bread. The peasants in the countryside wanted land to call their own. This gave rise to the countrywide calls for "Peace, Land, and Bread"!

Everyone, that is, *except* Tsar Nicholas Romanov! He had other ideas! And as he was the Tsar, which is to say Emperor, his power was almost absolute! That is a fact! It is also a fact that the Romanovs had ruled the vast Russian Empire for over *three hundred years!* Nicholas was an Emperor who was of the opinion that his subjects required a "firm hand"! He *earned* the title of "Nicholas the Bloody"!

Nicholas was aware of the rising tide of discontent that was sweeping his Empire, but not terribly concerned. After all, he had seen it all before! In fact, his ungrateful subjects had dared to mount a full scale revolution in 1905, and yet he had managed to weather the storm! He had managed before, he could manage again! Or so he thought!

In February of 1917, Tsar Nicholas was overthrown, within the space of eight days! The power of revolution!

How could an Empire, which had maintained itself for three hundred years, and recently survived a revolution, collapse in such a short time?

Lenin provides us with the answer.

At that time, Lenin was in exile, living in Switzerland, but closely following events in Russia. He was able to keep in touch with the Marxists, within Russia, by sending them a few letters. These are referred to as "Letters From Afar", and are most illuminating. This gives us an idea of just what can happen during a revolution! Nothing short of miraculous!

Lenin explained that there were a "number of factors", of "world historic importance", which were "required for the tsarist monarchy to have collapsed in a few days". Not the least of these was the fact that the Russian workers and peasants had gained valuable *experience* in the Russian Revolution of 1905-1907. As he stated, "The first revolution (1905) deeply ploughed the soil, uprooted age-old prejudices, awakened millions of workers and tens of millions of peasants to political life and political struggle...This first revolution, and the succeeding period of counter revolution (1907-14) laid bare the very essence of the tsarist monarchy... Without the Revolution of 1905-07, and the counter-revolution of 1907-14, there could not have been that clear 'self-determination' of all classes of the Russian people and of the nation's inhabiting Russia".

Note that here, Lenin stresses the fact that it was the *experience* the common people of Russia gained, *not only* in the first Revolution of 1905, but *also* in the *experience* of the counter revolution! That led to "clear self-determination"!

Lenin went on to explain the fact that the First World War acted as a "stage manager", one which "vastly accelerated the course of world history". As well, it engendered "worldwide crises of unparalleled intensity- economic, political, national and international".

It is important to note that the First World War did *not cause* the February Revolution, but *"vastly accelerated"* that event! It also gave rise to "worldwide crises of unparalleled intensity"! Such crises tend to give rise to revolution!

The February Revolution came as a complete shock, as no one could have expected the Tsar to be deposed, so quickly! The explanation was provided by Lenin.

As he stated, "That the revolution succeeded so quickly and -seemingly, at the first superficial glance- so radically, is only due to the fact that, as a result of an extremely unique historical situation, *absolutely dissimilar currents, absolutely heterogeneous* class interests, *absolutely contrary* political and social strivings have *merged,* and in a strikingly 'harmonious' manner". (italics by Lenin)

This is to say that the common people of Russia wanted Nicholas deposed, because they wanted *peace!* At the same time, the capitalists of Russia *also* wanted Nicholas deposed, but for precisely the *opposite* reason! The capitalists wanted *war! Same goal, opposite reasons!*

The Russian capitalists were well aware that the wife of Nicholas, was a German princess! Her title was Princess Alexandra von Hesse-Darmstadt! She wanted no part of war with Germany! The capitalists were afraid that her husband, Tsar Nicholas, would listen to her! The last thing they wanted, was for Nicholas to arrange a separate peace with Germany! The war was too profitable! Their British and French friends agreed with them! So "palms were greased" in the palace, high ranking government officials were bribed, and a "coup" was arranged! Nicholas was removed from power! Precisely what the common people wanted!

This, the First Russian Revolution of 1917, referred to as the February Revolution, served to *simplify* the class conflict! Tsar Nicholas was overthrown, so that the nobility was no longer in power. One reactionary class was out of the way! That left two other reactionary classes, the monopoly capitalists, the bourgeoisie, and the class of landlords.

The monopoly capitalists were anxious to *continue* the slaughter, that which has been referred to as the First World War, as it was producing such huge profits! The fact that the war was also ruining the country, was of no consequence. Merely the "cost of doing business"!

As well, the landlords owned all the land and were accustomed to "gouging" the peasants. Most of the crops the peasants grew, was handed over to the landlords, as rental payment. If this in turn led to the starvation of the peasants, that was of no concern to the landlords!

The vast majority of peasants were classified as "poor peasants", accustomed to working alone, plowing their fields, trying to grow enough crops to support their families. This rarely happened, to that in the winter, many of them would look for work elsewhere. In this way, they became "semi proletarians".

A few peasants were classified as "rich peasants", in that they were "successful", able to invest in machinery and hire labourers. They also took great delight in "gouging" their less fortunate neighbours, the poor peasants! They well earned the title "kulaks", meaning "tight fists", or "rural bourgeoisie"!

As can be well imagined, that left those classified as "middle peasants". They generally were able to manage reasonably well, on the crops they grew.

The middle peasants tended to correspond to the middle class, petty bourgeois. This is to say that they generally vacillated, in the class struggle, between the proletariat and the capitalists.

Even after the Tsar was overthrown, the revolution continued! The capitalists and landlords were of the opinion that the revolution had gone far enough! The common people, workers and peasants, were of an entirely different opinion! They thought that the revolution had *just begun!*

True, the Tsar was overthrown, so that the common people had certain democratic rights. Yet it is also true that you *cannot eat democratic rights!*

All common people remained hungry, cold and destitute! *All* peasants remained *crushed* by the landlords! *All* soldiers continued to suffer and die at the front! *All* common people of Russian, both workers and peasants, wanted "Peace, Land and Bread"! That became the slogan, the "battle cry"!

Sound familiar?

Now the suffering of the common people, in North America, is also at monstrous levels! Countless people are now unemployed, hungry, cold and homeless! Denied even the most basic medical care! Overdoses and suicides are commonplace! Criminal gangs now rule the streets! Mass murders are a daily event! The police are powerless against the heavily armed gangs!

The common people of North America are now almost entirely *working class! Proletarians!* The only *consistently revolutionary class!* In opposition to the completely *reactionary class,* the *billionaires,* the *bourgeoisie!*

As mentioned previously, Lenin said that the First World War acted as a "stage manager", which served to "vastly accelerate the course of world history". In much the same way, here in North America, the combination of the Corona Virus and the Second Great Depression, are also serving to "vastly accelerate the course of world history"!

Of course, government officials maintain that the Virus is under control, just as they claim that the banking system is under control. Such is hardly the case! The Virus is merely mutating! The recent collapse of three banks is merely the tip of the iceberg! The "cash infusion" of over *one hundred billion dollars* failed to save those banks! They were *Too Small To Succeed!* Countless others are sure to follow! That will necessarily lead to the collapse of countless small businesses, those with assets of *less than one trillion dollars!* In turn, the stock market is sure to collapse! The *Second Great Depression!*

As for the *experience* of revolution and counter revolution, it is true that Americans have not experienced a recent revolution, but they have experienced something very similar. In the nineteen sixties, countless people took part in various mass movements. In particular, the Viet Nam Anti War Movement was most impressive!

After the mass movements died down, those same people then felt the full wrath and fury of the counter-revolutionary forces! Years later, as the revolutionary motion picked up again, in the form of the Occupy Movement, a whole new generation gained valuable experience, in the class struggle.

We can think of this experience of revolutionary movement, as well as that of counter revolutionary motion, as that of "ploughing the soil", as "dispelling old prejudices", as "awakening millions of workers to political life".

This happened in Russia, in the early twentieth century. Now, it is taking place in North America.

Those American working people, young and not so young, who took part in those mass movements, are now seasoned veterans! Schooled in the class struggle! They know what to expect! They are now prepared for full scale revolution!

Working people are not about to put up with the current situation! These "seasoned veterans" are now "expanding their horizons", becoming class conscious -*revolutionary!*- by reading State and Revolution!

In Russia, it was first necessary to raise the level of awareness, of the common people. Only then could a successful insurrection take place. Here too, in North America, it is necessary to raise the level of awareness of the working class. After all, there are a great many working people who are still not convinced of the necessity of revolution. So many still believe the lies of the billionaires, to the effect that democracy means "majority rule". We must respect their belief!

As is well known, I have previously made a number of suggestions, with a view to raising the level of awareness of the working people. It is important that all common people become politically active! With a little experience, in the class struggle, they will soon come to believe that the

Marxists are correct! The billionaires *must* be overthrown, and this can *only* happen through *revolution!*

In the interest of gaining personal experience, I suggested that all common people join the two mainstream political parties, as card carrying members. I also suggested that those same people get involved with Councils.

I further suggested that all workers read State and Revolution. A careful reading of that classic work by Lenin, will provide working people with a fine understanding of the revolutionary theories of Marx. This is to say that all workers should combine experience with revolutionary theory.

And now I have another suggestion! In the interests of *raising the level of awareness* of the working class, by taking an *active role in the class struggle, of securing their democratic rights*, may I suggest that all Americans first read the Twelfth Amendment to the Constitution. It lays out the procedure to be followed in all *federal elections!* Here is a summary of the key parts:

"The Electors shall meet in their respective states and vote by ballot for President and Vice President…they shall name in their ballots the person voted for as President and in distinct ballots, the person voted for as Vice President.."

This procedure was *not* followed in the 2020 federal election! As I have documented this in a previous article, there is no need to repeat it here. Suffice it to say that it is grounds for *challenging that election,* in a court of law! If that happens, it will very likely make its way to the Supreme Court, very quickly.

Should the Supreme Court rule that the 2020 federal election was *fraudulent,* then it follows that Biden is a fraudulent President, and Harris is a fraudulent Vice President. Grounds for removal from office!

The Supreme Court could also rule that the *next* federal election, that of 2024, *must* follow the procedure laid out in the Twelfth Amendment. The implications are that the billionaires have every reason to be concerned!

Democracy is under attack! The democracy of the billionaires! Their method of class rule! It is Unconstitutional!

Any popular vote is meaningless! The only votes that count, is that of the Electors! It is only the states that can appoint Electors! The Electors are free to vote for the individual of their choice for President! The Electors are free to vote for the individual of their choice for Vice President! Such individuals may or may not be candidates of a particular political party! Such individuals may not even be a member of a mainstream political party! Electors choice! All state laws, requiring Electors to vote for any particular candidate, will be struck down as Unconstitutional!

The states do *not* have the right to force *any Elector* to vote for the candidate of *any political party!* For President or for Vice President! The Electors are free to vote for the candidates of *their choice! That is the law! Constitutional law!*

The popular vote is a mere *formality!* The particular candidate of either mainstream political party, for President, or for Vice President, is meaningless!

Trump maintains that Biden is a *fraudulent President!* He also maintains that Harris is a *fraudulent Vice President! Which they are!* But only because the 2020 federal election was *fraudulent! As were all federal elections for about the last one hundred fifty years!*

A ruling of the Supreme Court, to the effect that the 2020 federal election, was indeed *fraudulent,* would prove that Trump is *correct!* In the sense that the 2020 federal election *was* fraudulent, but *not* for the reasons that Trump maintains! It was fraudulent because it was *Unconstitutional!*

It is very likely that Trump, and his most devoted followers, will *embrace* such a challenge, to the 2020 federal election! *Not* that he is concerned with "free and fair elections", but *only* because he wants to be President! He may even help to pay for the court expenses! He would love nothing better than to have Biden and Harris removed from the White House! Always happy to help out a worthy cause!

Trump is determined that he won the 2020 federal election! His argument is that he "won the popular vote", because the Electors should have voted for him, as *required* by state laws! Yet those state laws, which *require* an Elector to vote for an individual who is a candidate of a particular mainstream political party, are *Unconstitutional!* The states have *no right* to meddle in a federal election!

Such a ruling, by the Supreme Court, that of striking down all state laws requiring the Electors to vote for particular candidates, is not about to impress Trump! He is not a man who is terribly impressed by logic! He will still argue that he is the true President! Which he is *not!* In fact, America has *not* had a true President, or true Vice President, since the days of the Civil War!

The billionaires will oppose any challenge to the 2020 federal election, as they consider it to be a threat to their *method of rule!* Just as any demand that the 2024 federal election be conducted according to the procedures laid out in the Twelfth Amendment, is *also a threat to their method of class rule! Their democracy!*

Any Supreme Court challenge to the 2020 presidential election, by the working people, is sure to be welcomed by Trump! He may even welcome a demand that the 2024 federal election follow the procedures laid out in the Twelfth Amendment!

At the same time, the billionaires would oppose any such challenge! The billionaires see such a challenge as a threat to their *method of class rule!* They refer to this method of rule as *democracy!*

This challenge, to the Supreme Court, *could* happen! Which is *not* to say that there would be an "alliance" of Trump with the working class! By no means! It is to say that this would be an example of the *"merger"* of the two *goals* of two *class enemies!*

The billionaires are desperate to make sure that Trump, who is *one of their own,* but a *renegade, is not* once again elected President! They see Trump as a threat to their democracy! Which he is! Trump is definitely a *threat to their democracy!* Bear in mind that *their democracy* is merely their *method of class rule!*

Trump plans to once again become President, and set himself up as *President for Life!* That does *not* fit in with the current system of *democracy,* the current *method of class rule,* of the billionaires! If Trump is successful, then the billionaires will be forced to *change their method of rule!*

This is to say that we have the progressive working class, the proletariat, defending their democratic rights. Challenging the 2020 federal election as Unconstitutional! Demanding instead, a 2024 federal election that follows the procedure laid out in the Twelfth Amendment!

At the same time, we have the forces of reaction, led by Trump, a *renegade* billionaire, who is also anxious to challenge the 2020 federal election, for precisely the opposite reason! He wants to be President! Not only that, but President for Life!

This is another example of that which Lenin referred to as "absolutely dissimilar currents", completely contrary "class interests", coming together, "merging", in a "harmonious manner"!

This happened in Russia, in early 1917, and it could happen in America!

The working class of America, or at least the more advanced workers, want to secure their democratic rights, as is guaranteed in the Constitution. That includes a free and fair federal election, as per the Twelfth Amendment. For that reason, they are prepared to challenge the 2020 federal election, preferably in the Supreme Court. They are demanding a free and fair, *Constitutional 2024 federal election!*

Yet just because Trump *welcomes* any Supreme Court challenge, to the 2020 federal election, does *not* mean that the working class should *oppose* that challenge! On the contrary, there is a *principle* involved! Working people have a *duty* to "preserve, protect and defend" their democratic rights, as guaranteed in the Constitution! If their democratically elected leaders will *not* perform their duty -and they will *not!*- then the working people *must perform this duty for them!*

It is in the *process* of challenging the 2020 federal election, while at the same time *demanding* a free and fair 2024 federal election, in a court of law, that the American *working class* will gain valuable *experience* in the *class struggle!* Their "level of awareness" will be raised! They will learn, from *experience,* that the billionaires are in charge, and fully intend to remain in charge!

In the case of Russia, in 1917, this "merger" of the goals of the capitalists and landlords, with the goals of the workers and peasants, resulted in the "First Revolution" of February, 1917. The Tsar was overthrown, and the common people had some democratic rights, at least on paper.

If nothing else, it allowed Lenin to return from exile. This he did, in April of 1917. The *first stage* of the revolution had been completed. The autocracy, in the form of Tsar Nicholas, had been overthrown. That merely placed the capitalists and landlords in charge. The government they formed was known as the Kerensky Provisional Government.

The *second stage,* that of *overthrowing* the capitalists and landlords, and establishing a socialist republic, was next on the agenda! Lenin wasted no time in *preparing* the common people of Russia, for the approaching revolution!

This is to say that Lenin was focused on *raising the level of awareness* of the common people, the workers and peasants! *Especially the working class, the proletariat!* Because it is the proletariat that is the only *consistently revolutionary class!* The proletariat is the class that will *lead the revolution!* The peasants and middle class will follow along!

Lenin was well aware that the working class is not aware of itself, *as a class,* with its own *class interests!* It follows that they are equally unaware of the fact that the capitalists, the bourgeoisie, also exist, *as a class!* They too, have their own *class interests!* It is in the interests of the capitalists to force their workers to work as hard as possible, while paying them as little as possible! That is *not* in the best interests of the workers! The interests of the two classes are *diametrically opposed!* For that reason, the capitalists have to be *overthrown!* This can only happen through *revolution!*

With that in mind, in the interests of making the working class aware of this, that Lenin wrote State and Revolution. This may well be considered to be a "road map" to revolution!

In State and Revolution, Lenin first explained that the "state is the product and the manifestation of the *irreconcilability* of class antagonisms."(italics by Lenin)

(For the benefit of those readers who are not Rhodes scholars, I will mention that "irreconcilable" just means "impossible to find agreement between".)

This is to say that the mere fact that the *state exists*, is proof that *class antagonisms,* between the workers and the capitalists, *cannot* be resolved, at least not through negotiations! Further, the existing state apparatus has been carefully set up, *by the billionaires*, for the express purpose of *crushing* the working class! For that reason, it must be *smashed*, at the time of the revolution!

This brings us to the social chauvinists. They *claim* to be Marxists, while maintaining that the revolutionary theories of Marx and Lenin should be *revised!* They want *no* part of smashing the existing state apparatus! They have plans for that state apparatus! They plan to *take control* of that state apparatus, at the time of the Revolution. and use it for *their own purposes!* They plan to use it to *set themselves up as the new rulers!*

As a means of stressing the importance of *smashing* the existing state apparatus, Lenin went on to say, "if the state is the product of irreconcilable class antagonisms, if it is a power standing *above society* and '*increasingly alienating itself from it*', it is clear that the liberation of the oppressed class is impossible, not only without violent revolution, *but also without the destruction* of the apparatus of state which was created by the ruling class". (italics by Lenin)

Strangely enough, there are times when the state apparatus becomes something more than just an instrument of class rule! As Engels observed, "By way of exception, however, periods occur when the warring classes are so nearly balanced that the state power, ostensibly appearing as a mediator, acquires, for the moment, a certain independence in relation to both".

This is significant, for a couple of reasons. Historians have often wondered the reason that the agents of the Kerensky Regime did *not* arrest Lenin, as soon as he stepped off the train, in Saint Petersburg, upon his return from exile. Because "Soviet Power" was too strong! They dared not arrest Lenin! Even by April of 1917, Soviet Power was a rival to the government power!

As I have mentioned in a previous article, Soviet Power first appeared, quite spontaneously, in Russia, at the time of the First Russian Revolution of 1905. As that Revolution was crushed, in 1907, Soviets were also crushed. Yet as the revolutionary motion picked up again, in 1914, the Soviets once again, made an appearance. It is clear that revolutionary motion frequently gives birth to Soviets.

It is also clear that, within a rather short time, these Soviets became strong enough to rival the power of the Provisional Government! This is just one more indication of the power of revolution!

I mention this because in America, these Soviets have also made an appearance. They are referred to as Councils. For the moment, they may be weak, but soon they will challenge the power of the federal government! They are not to be under estimated!

Now to return to State and Revolution.

Lenin goes on to explain that after the existing state apparatus of the capitalists is *destroyed*, it is still necessary to *crush* the capitalists, as they make every effort to "regain their paradise lost".

As Lenin stated, "The overthrow of bourgeois rule can be accomplished only by the proletariat, as the particular class whose economic conditions of existence train it for this task and provide it with the opportunity and the power to perform it. While the bourgeoisie breaks up and disintegrates the peasantry and all the petty bourgeois strata, it welds together, unites and organizes the proletariat. Only the proletariat -by virtue of the economic role it plays in large scale production- is capable of acting as the leader of *all* the toiling and exploited masses, whom the bourgeoisie exploits, oppresses and crushes not less, and often more, than it does the proletarians, but who are incapable of waging an *independent* struggle for their emancipation….The overthrow of the bourgeoisie can be achieved only by the proletariat becoming transformed into the *ruling class,* capable of crushing the inevitable and desperate resistance of the bourgeoisie, and of organizing *all* the toiling and exploited masses for the new economic order". (italics by Lenin)

A careful reading of State and Revolution, by the more advanced members of the working class, will go a long ways towards raising their level of awareness! They are about to become *revolutionary! True Communists!*

This is excellent, but in no way changes the fact that working class people need leaders! A *true Communist Party is still needed!* The only true Communist Party is one which calls for the Dictatorship of the Proletariat!

As Lenin went on to explain, "Marxism educates the vanguard of the proletariat which is capable of assuming power and of *leading the whole people* to socialism, of directing and organizing the new order, of being the teacher, guide and leader of all the toiling and exploited in the task of building up their social life without the bourgeoisie and against the bourgeoisie". (italics by Lenin)

This is the task of the Communist Party! The creation of such a Party is very likely *beyond* the ability of almost all workers! It is certainly *within* the ability of well-educated middle class intellectuals! Equally without doubt, certain of the most advanced workers can help to create such a Party! This is now in the hands of middle class intellectuals!

I am convinced that the current situation, in America, is similar to the situation in Russia, immediately after the February Revolution. At that time, the tension was extreme, as it is now!

As a result of the explosive situation in Russia, in July of that year, there was a *spontaneous* uprising! Countless working people rose up, especially in the capitol of Saint Petersburg. The response of Lenin was instructive! He called for calm!

This uprising has gone down in history as the "Revolutionary July Days". It did *not* lead to an insurrection, because the working people took the advice of Lenin, and calmed down!

In the summer of 1917, Lenin was convinced that the vast majority of common people, the workers and peasants, were *not yet* convinced of the *necessity* of revolution! It was very likely that an insurrection, *at that time*, would have failed! It was too soon! It did not have the support of the vast majority of working people! The level of awareness of the common people had to be raised! They had to become *class conscious!* In the stilted jargon of the capitalists, "public opinion" had to become *revolutionary!*

For that reason, after the uprising had calmed down, Lenin ordered the members of his Party to go among the working people, and *raise their level of awareness!* They had to become *convinced of the necessity of revolution!* Under exceptionally difficult circumstances, against all the odds, they succeeded! Within a very short time - several months!- the common people were prepared for the Socialist Revolution! This has gone down in history as the Great October Socialist Revolution!

In much the same way, I am convinced that the American working class people are *not yet* ready for a socialist revolution! There are still a great many working people who are not convinced of the *necessity* of revolution. They still believe the lies of the billionaires. These people must be *respected!*

We can only urge these people to become politically active. Join the two mainstream political parties! Change the system from within! Challenge the capitalists in court! Demand that the 2020 federal election be declared fraudulent! Demand that the 2024 federal election follow the procedure laid out in the Twelfth Amendment! Demand fair and open elections! Hold your democratically elected politicians accountable! Become involved with the Councils that are appearing across the country! And by all means, read State and Revolution! Gain *experience* as well as *knowledge* in the *class struggle!*

It is only in this way, by *combining* experience with theory, that the working people of America will become *class conscious!* This is to say that the working people will become aware that it is *necessary* to overthrow the billionaires, *smash* the existing state apparatus, and *replace* it with the *Dictatorship of the Proletariat!* It is necessary that working class people become *Marxist Revolutionaries! Communists!*

Without a doubt, that is a "tall order". But to borrow a poker player expression, we have to "play the hand we have been dealt". We do not have to like it. We just have to do it!

Yet a spontaneous uprising could happen, at any time! Any "spark" could cause an explosion! An uprising!

By all means, Councils should continue to assist the desperately poor, in every way possible. As well, they should continue to arm, equip and train workers, in preparation for the insurrection. Also, get in touch with other Councils, in preparation for a country wide insurrection. The revolution will happen when it happens! Certain things are simply beyond human control!

Then too, everyone should join the two mainstream political parties, preferably as card carrying members. Make every effort to "change the system from within". Use this as an opportunity to raise the level of awareness of the working class. Encourage workers to become politically active, to learn *from their own experience!*

Experts on Constitutional law should challenge the 2020 federal election, in court. Focus on *defending* the Constitutional rights of all Americans! Demand a *legal 2024 federal election! One which follows the guidelines of the Twelfth Amendment!* Use this as a political platform! No doubt, countless working people will be watching this court challenge! In this way, their level of awareness will be raised!

Everyone should be encouraged to read State and Revolution! Bear in mind that, during a time of revolution, countless working people, including those who were previously apathetic, become *politically active!*

The distinction, between the "most advanced" and the "less advanced", becomes less obvious!

This brings us to a statement by Lenin, in Left Wing Communism, An Infantile Disorder. As I have previously quoted the paragraph in full, in an earlier chapter, I will merely quote the essential parts.

"Symptomatic of any genuine revolution is a rapid, tenfold and even hundredfold increase in the size of the working and oppressed masses -hitherto apathetic- who are capable of waging the political struggle".

That is an accurate and detailed description of our current situation! We are indeed in the midst of a "genuine revolution"! Countless working people, who were "hitherto apathetic", are now "waging the political struggle"! This serves to blur the distinction between the most advanced workers, and the less advanced! *All* must be encouraged to become politically active! *All* must be encouraged to read State and Revolution!

Those who are politically active must *flood* social media with calls for *revolution, Soviet Power and the Dictatorship of the Proletariat!* All marches and demonstrations must also make such calls!

Those who are involved in the entertainment industry, must find creative ways to spread the revolutionary message! Possibly through song and dance! Bear in mind that the best way to *educate* people, is to *entertain* them, at the same time!

By all means, middle class intellectuals should take part in the creation of a true American Communist Party, Dictatorship of the Proletariat, ACP,DP. Be discrete! Avoid the phone! Use the internet! Preferably the "dark net", whatever that is! Make the "spooks" in the government spy agencies earn their pay!

Bear in mind that, in 1917 Russia, under far more difficult conditions, the working people became revolutionary within a matter of a *few months!* Now in America, with the working class so cultured, complete with the internet, it is *so much easier* to raise their level of awareness!

Now to return to the subject of Russia, 1917.

It was only *after* the level of awareness of the working people had been raised, that Lenin decided it was time for the Insurrection. The Second Russian Revolution of 1917, took place on October 25, old style calendar, or November 7, new style calendar. It has gone down in history as the Great Socialist October Revolution. That revolution was completely successful, and almost bloodless.

We must stress that this revolution was successful, only because it *followed the advice of Lenin!* It was based on *scientific principles!* The revolution established a society of *scientific socialism!* The one and only true form of socialism! The *Dictatorship of the Proletariat! Experience* has shown that all other forms of socialism simply *do not work!*

In regards to our approaching American Revolution, the situation is much simpler, as the nobility has long since been "given the boot". As well, there is no longer a class of landlords to contend with, or the corresponding class of peasants, farmers. Granted, the remnants of those classes are still in existence. In much the same way, the middle class is in the process of being wiped out.

Simple the task may be, but that is *not* to say that the task is easy! Indeed not! In fact, with their huge profits, the billionaires have managed to "grease the palms", to *bribe*, a great many members of the working class!

Lenin documents this quite well in Imperialism, the Highest Stage of Capitalism. He points out that the imperialists, the billionaires, are able to "corrupt the upper strata of the proletariat", with some of the proceeds of their huge profits.

In the case of America, it is clear that the "corrupted upper strata of the proletariat", is strictly "gender specific"! Not too many American women have been "corrupted" by those bribes! If those taking part in the Me Too Movement are any indication, such "corrupted" women are few and far between! Most American women are completely fed up! What is more, the women are *leading* the American Revolution!

Now to return to the Great October Socialist Revolution. At that time, we must stress that the common people followed the "road map", as laid out by Lenin, in State and Revolution. The existing state apparatus was promptly *smashed!* A new state apparatus was then established, in order to "crush" the "desperate and determined resistance of the capitalists", as they tried to "restore their paradise lost". This state apparatus is referred to as the Dictatorship of the Proletariat!

Immediately, a number of republics, which had been crushed under the Russian Empire, declared independence. Among these newly created independent republics, were several which also established socialist societies.

At first, none of these independent republics completely trusted the Russian Soviet Socialist Republic. They had been crushed by Russia for so many years! They had no reason to trust any Russian! Yet after several years, they became convinced that Lenin and the other high ranking Marxists, within the Central Committee of the Communist Party, meant precisely what they said! At that point, they joined with Russia, to form the Union of Soviet Socialist Republics.

We can expect a similar situation to play out in America. As the billionaires are overthrown, no doubt several republics will declare independence. It is very likely that most of them, if not all, will declare themselves to be socialist republics!

The implications of this are also obvious. The burden on the people who are leading this, the Second American Revolution, just became greater. They not only have to plan a nationwide insurrection, but also plan to establish a number of independent socialist republics, immediately after the successful insurrection. Further, it is quite likely that the revolution will cross borders, into Canada and Mexico, possibly further.

Lest people scoff, bear in mind that New England has recently broken in two. The southern three states of Massachusetts, Connecticut and Rhode Island, have joined with the states of New York, New Jersey, Pennsylvania and Delaware. The northern three states, that of Maine, Vermont and New Hampshire, could very well join with the Canadian

Maritime Provinces, that of New Brunswick, Nova Scotia and Prince Edward Island. Each could form an independent socialist republic. They have a great deal in common.

As well, seven states in the midwest, the industrial heartland of the country, have just come together. Then too, three highly industrialized states on the west coast have also just combined. Each will almost certainly form an independent socialist republic.

Further, Alaska could separate and join with the Canadian Yukon, for example. Another possible independent socialist republic! These are merely several examples, and no doubt, the people who live in those areas, as well as a great many others, will decide their own future.

With the possible exception of the Great October Socialist Revolution, never before has a revolution on this scale taken shape! Yet never before has the working class, the proletariat, been so schooled in the class struggle! So many of them are now seasoned veterans, ready to take up arms against their class enemies, the billionaires. The only thing lacking now is the leaders!

With that in mind, my message, to so many former members of the middle class, is that I suggest you cast aside your illusions. Your previous life style is gone, never to return! Your years of faithful service, to the billionaires, count for nothing! Expect no reward! Do not look back! Forward!

Focus now on building a new society, a socialist society! Your knowledge and skills are needed! After the revolution, you will be in demand! You will be paid accordingly, just as all professional people will be well paid. Do not get mad! Get even!

Take part in the creation of a proper American Communist Party, Dictatorship of the Proletariat. Bring to the working class the consciousness they so desperately need. Take a leading role in the revolution. You have a bright future under socialism!

We will know that our message is getting through when the banners and posters read:

Scientific Socialism!

Dictatorship Of the Proletariat!

Workers of the World, Unite!

CHAPTER 9

CAPITALIST OPPOSITION TO INFRASTRUCTURE REPAIR

Recently in Florida, a twelve story apartment building collapsed, killing a great many people. The press is covering this story very closely, properly so. They are also reporting that this collapse raises a great many questions concerning the integrity of other buildings, as well as the infrastructure of the whole country. In this they are correct, so perhaps it is best to explain these terms.

The building in question was over forty years old and constructed with concrete, which was reinforced with steel bar. That steel bar is referred to as rebar, which is short for reinforced bar. This rebar gives the concrete more strength. The press also reports that the concrete on the building was "spalling", or cracking, and that this cracking is an indication that the steel rebar is rusting.

Of course the collapse of this building raises the question of the safety of other buildings, especially those which are greater than forty years old. It also raises the question of the safety of the "infrastructure" of the whole country. The press was also quick to point this out, without explaining the meaning of the term infrastructure.

Perhaps someone should explain to the journalists, that not everyone who watches the news is a cross between a Philadelphia lawyer and a mechanical engineer. With that in mind, we will mention that, according to the internet, the infrastructure is a reference to "the basic physical and organizational structures and facilities, such as bridges, roads, power supplies and tunnels, needed for the operation of a society." They go on to mention that this includes the water and sewage network. Now we know.

This is certainly a legitimate concern, as many of those structures were built a great many years ago. It stands to reason that without proper routine maintenance, it is just a matter of time before they fall apart. Yet maintenance costs money, and the capitalists have been cutting back on such expenses for many years. It follows that a great many more buildings are about to collapse, and not just apartment complexes.

It only makes sense to repair the infrastructure of the country, before it falls apart, in the interest of keeping the country running smoothly, as well as for the sake of safety. This would also help to put countless people to work, performing vital repairs. In the process, those workers will be sure to be paid, quite handsomely. This would help to stimulate the economy, and in fact, Senator Bernie Sanders has recently proposed a *six trillion* dollar stimulus package, for those repairs. Yet the capitalists are opposed to this!

Recently, in one of the most highly respected publications of the capitalists (respected by the capitalists), they revealed their thought processes, to put it politely. To put it more accurately, they revealed their absolute stupidity and total disregard for human life! As they phrased it:

"Crises have a tendency to kill old orthodoxies and usher in new ones. The horrors of the Great Depression galvanized governments to start fighting recessions, instead of waiting for the market to work itself out...*It is better to overdo it on stimulus than to under do it,* economists concluded, *next time we'll prime the economy with the mother of all pumps".* (italics by the capitalists)

The article went on to sing the praises of perhaps the finest of the bourgeois economists, John Maynard Keynes. He was active in the early twentieth century. Yet even though he was admired by the capitalists, at no time did they ever take his advice!

At the end of the First World War, it was Keynes who advised the Allies to cancel all debts, and to return to the borders that existed, before the start of the war. They most certainly did not!Instead, at the Treaty of Versailles, the Allies imposed staggering war reparations on Germany. As Keynes put it, this was sure to lead to "serious economic and political repercussions on Europe and the world". He was so right!

Now to return to the same article:

"That's exactly what happened when the greatest crisis of the century (the twenty first century) came crashing down. Congress approved trillions of dollars in Covid relief...It was the kind of government activism that would have stunned even John Maynard Keynes...A little over a year into the crisis, the preliminary results of those actions are in, and they're phenomenal. The US economy is on track this year to grow at its fastest pace since 1984...they could put millions of Americans to work.... It's Keynes on steroids"

So far so good. The capitalists finally took the advice of their own bourgeois economist, an expert, and were amazed by the results. But now comes the kicker:

"But... A growing chorus of voices ... fear that government spending *...would enable the workers to demand higher wages and force the wealthy to pay a fairer share of taxes".* (my italics)

There we have it! The capitalists are afraid that if they invest too much money in infrastructure maintenance and repairs, then the standard of living of the working class will rise, the working class will become ever stronger, they will force the capitalists to increase the wages, and even force the billionaires to "pay a fairer share of taxes". Better to allow the infrastructure of the country to fall apart, rather than pay the workers higher wages, as well as having the billionaires pay higher taxes!

Without doubt, the capitalists, and especially the bourgeois economists, are supremely well aware of the theories of Marx. They even admit, in private, that Marx was correct. They know that Marx was right, when he said that "reforms are a byproduct of revolutionary motion...which serve to strengthen and further the revolutionary motion". All the more reason to *oppose* those reforms! Even if that means running the country into the ground! Better to allow the infrastructure of the country to fall apart, rather than force the billionaires to pay taxes!

We live under monopoly capitalism, which is imperialism. As Lenin pointed out, imperialism is completely reactionary. Billionaires are monopoly capitalists. Billionaires are imperialists. Billionaires are reactionaries. This is just one more example of such reaction.

As stated earlier, our civilization has passed its peak, and is now in decline. Yet our civilization has experienced the one thing that all preceding civilizations have not! An industrial revolution! This gave birth to a truly revolutionary class! The proletariat!

It is the proletariat that will reverse this decline in our civilization! The proletariat is destined to overthrow the billionaires, smash the existing state apparatus, and set up a new socialist state, in the form of the Dictatorship of the Proletariat! It is the proletariat that will make sure that our civilization does not collapse, as all previous civilizations have collapsed!

Onwards, fellow revolutionaries, to the Dictatorship of the Proletariat!

CHAPTER 10

SUPPORT BERNIES SANDERS

There is only one Senator who is calling for a massive improvement to the infrastructure of the country, and that is the Independent Senator from the state of Vermont, the self-described Independent Socialist, Bernie Sanders. He is now Chairman of the powerful Senate Budget Committee, and is not satisfied with the *3.5 trillion dollar* infrastructure package in domestic investments. Sanders is arguing that *6 trillion dollars* is required!

It is clear that Senator Sanders is a devoted follower of John Maynard Keynes, one of the most highly respected bourgeois economists of the twentieth century. As the current bourgeois economists phrase it, "the main plank of Keynes's theory, which has come to bear his name, is the assertion that aggregate demand -measured as the sum of spending by households, businesses and the government- is the most important driving force of an economy".

To think that the bourgeois economist who came up with that "plank", is considered to be one of the best and brightest! What else could "demand" be, but the "sum of spending of households, businesses and government"?

And how could that not be the "most important driving force of the economy"? Yet Keynes is considered to be a genius for pointing this out!

Perhaps it would be best to allow Sanders to explain, in his own words, as he expressed himself recently, after meeting with President Biden: "He knows and I know that we're seeing an economy where the very, very rich are getting richer, while working families are struggling...My job is to do everything I can to see that the Senate comes forward with the strongest possible legislation to protect the needs of the working families of this country... We want to see a reconciliation bill which shows the working families of this country that the government can and must work for them.. What we are trying to do is transformative. The legislation that the president and I are supporting will go further to improve the lives of working people than any legislation since the 1930's... Does anyone deny that our child care system, for example, is a disaster? Does anyone deny that pre K, similarly, is totally inadequate? Does anyone deny that there's something absurd that our young people can't afford to go to college, or are leaving school deeply in debt? Does anyone deny that our infrastructure is collapsing?"

Well spoken, Bernie Sanders! No one can deny that!

It is clear that Sanders is one of the finest "Independent Socialists", technically referred to as a utopian socialist. He has stated the problem clearly, while carefully avoiding any reference to classes. The "very, very rich" is a reference to the billionaires, and he is correct that they are "getting richer". The "working families" is a reference to the working class, the proletariat, and it is true that the working class is "struggling". Or at best, many of the working people are struggling, while so many others are unemployed, homeless, hungry and desperate. In fact, these people are not so much struggling, as being degraded!

As a result of the revolutionary motion that is sweeping the country, Senator Sanders has been placed in charge of the powerful Senate Budget Committee. As such, he is correct when he states that it is his duty to "protect the needs of the working families of this country". For that reason, he is proposing a massive "reconciliation bill", with the idea of providing proper education, from pre-kindergarten through college, along with a major overhaul of the infrastructure, among other things. What is more, he thinks that President Biden supports him on this. Now where would he get an idea like that? Probably from President Biden! As a means of stalling Sanders!

The fact of the matter is that many of the finest intellectuals among the capitalists, especially the economists, are well aware of the revolutionary theories of Marx and Lenin. What is more, they are also well aware that those theories are correct! All the more reason to do everything in their power to divert the revolutionary movement!

Those same bourgeois economists are well aware that, as Marx stated, "reforms serve to strengthen and further the revolutionary motion". For that reason, they are opposed to reforms! They tried to block the recent "economic stimulation package", in which almost all Americans received a "stimie check", and it worked wonders for the economy. It also served to "strengthen and further the revolutionary motion", as the bourgeois economists are well aware. They are afraid that the proposed "reconciliation bill" will do even more "damage", to the extent of workers demanding higher wages, as well as forcing the billionaires to pay a higher rate of taxes. God forbid!

Yet the fact remains that the "stimie checks" have served to strengthen the working class revolutionary motion, and it is time to build upon that. The best way to do that is by having all "Leftist" people come together, in support of Senator Sanders. It matters not if they consider themselves to be Independent Socialists, Democratic Socialists, Social Democrats, Communists or just plain Socialists. For that matter, they can be anarchists or people who are determined to see some change for the better! We can all agree to support Sanders, in his quest to force through the "reconciliation bill"!

Perhaps the best way to do this is to join the two mainstream political parties, both Democrat and Republican, as card carrying members. As such members decide the candidates for all political offices, they have great power. They can force elected officials to vote in favour of Sanders and his reconciliation bill, because if they do not, they will not be allowed to run for reelection. Faced with the prospect of career suicide, watch how fast they become sweetly reasonable!

In addition, middle class lawyers, especially those who are experts on Constitutional law, can challenge the 2020 federal election, on the grounds that it did not follow the procedures laid out in the Twelfth Amendment. They can also demand that the upcoming 2024 federal election follow those procedures. No doubt the working people will watch that court challenge very closely!

At the same time, the true Communists, those who call for the Dictatorship of the Proletariat, can use that platform to make clear that the problem is one of capitalism! The capitalists, the billionaires, must be overthrown, the existing state apparatus smashed, and the billionaires crushed, under the Dictatorship of the Proletariat. In this way, we can distinguish ourselves from the social chauvinists and the utopian socialists.

Bear in mind that our goal is to *raise the level of awareness* of the working class. The working class must be *made aware* of the fact that the *capitalist class*, the bourgeoisie, currently referred to as billionaires, must be *overthrown,* through *revolution,* and *crushed,* under the *Dictatorship of the Proletariat!*

With that in mind, we can also encourage the more advanced workers to read that most relevant work of Lenin, State and Revolution.

We are suggesting this as a course of action, in conjunction with the creation of a Communist Party, Dictatorship of the Proletariat. As well, the members of the various Zones, which have sprung up across the country, should prepare for the insurrection. That includes learning the proper use of firearms, especially twenty twos, as they are the least expensive. As well, both men and women should learn to use night sticks, pepper spray, paint balls and slings, complete with marbles. Become comfortable with the wearing of helmets and bullet proof vests. Each Zone should be provided with several pipe wrenches, and all should be taught to use them. Opening up a fire hydrant is simplicity itself. As the water trucks get their water from the fire hydrants, a simple way to neutralize them, is to open up all fire hydrants. In that case, the water pressure drops to zero, and the water trucks are deprived of water.

As the American women are currently leading the revolutionary motion, this falls mainly upon their shoulders. The women have proven themselves to be excellent organizers, and those are precisely the skills we need now.

It was Lenin who consistently stressed the importance of the "key link". The key link now is to prepare for Soviet Power and the Dictatorship of the Proletariat. Ladies, we have complete confidence in you!

CHAPTER 11

LADIES: ANARCHY OR SOCIALISM, YOUR CHOICE!

The current crisis in capitalism, which has been intensified by the Corona Virus, has led to widespread unemployment, homelessness, hunger, drug abuse, overdoses, suicides, and gang terror. On average, there are 54 people who are shot and killed in America *each day!*

In fact, that which is politely referred to as "mass casualty events", in which more than one person is killed or wounded, are now almost routine! It is just a matter of time before the system breaks down completely. At some point, both the medical and the law enforcement agencies will collapse. At that time, the country will be facing anarchy. No law. Survival of the fittest. Gangs of people with guns will be roaming at will, raping and looting as they please.

As for those who think that I am overstating the case, feel free to face the fact that such events have happened before. At the time of the collapse of a civilization, a period of anarchy always takes place. The American civilization is close to collapse.

It is clear that the government leaders in Washington, our democratically elected officials, are not losing any sleep over such little details. They are preoccupied with the "Big Lie", as well as the "January 6 Insurrection". The two are related, two sides of the same coin. The Big Lie is a none too subtle reference to the belief, among millions of Americans, that Trump won the presidential election, so that Biden is an imposter president. The demonstration in the capital of Washington, on January 6, in which protesters occupied the government building of the Congress, is being referred to as an "Insurrection". Those protesters are being referred to as anarchists, which they are not. They are devoted followers of Trump, convinced that he is the true president, and want him placed back in the White House.

The bourgeois politicians have their priorities well established. The death of so many Americans, due to the Corona Virus, now well over 600,000 and rising, as well as the shooting deaths of thousands more, does not concern them. Nor are they concerned with the widespread suffering of the working class! Their only concern is that of maintaining the power of the billionaires!

The fact is that the 2020 *federal* election was fraudulent, so that we currently have a fraudulent President, as well as a fraudulent Vice President. Which is *not* to say that Trump won the election! This is just to say that the process to be followed, in a federal election, as is outlined in the Twelfth Amendment to the Constitution, was *not* followed!

As we have gone into this in another article, there is no need to repeat it. Suffice it to say that it is *not* the citizens who elect the president, but the members of the Electoral College. The popular vote is a mere formality! Further, the states have no right to meddle in a federal election.

It is significant that the manner in which all elections can be falsified, was documented, quite clearly, by the former personal lawyer for Trump, Rudy Giuliani. It was a valuable public service, and revealed the fraud which takes place, on a regular basis, during so many elections. As a "reward" for this public service, he is now no longer allowed to practice law in the state of New York.

As previously mentioned, my suggestion is that an attorney, preferably an expert on Constitutional law, challenge the 2020 federal election, on the grounds that it did not follow the procedures laid out in the Twelfth Amendment to the Constitution. The argument can be made that the District of Columbia is not a state, so had no right to appoint Electors. As well, the argument can be made that the states have no right to meddle in a federal election. A different section of the Constitution can be referred to, in support of that particular argument.

The attorney can then request the Supreme Court to declare the 2020 federal election to be fraudulent. She can also request that the next federal election follow the procedures laid out in the Twelfth Amendment.

It is very likely that Trump will support this challenge to the 2020 federal election. After all, he is determined that the election was fraudulent! Yet if the Supreme Court rules that the election was indeed fraudulent, then it follows that the 2016 election, in which he was elected President, was also fraudulent!

Such details do not impress Trump! He is convinced that he is the true President! He is not a man to be swayed by logic! Clearly, he is not the "sharpest tool in the shed"!

Incidentally, I trust that no common people, who live in the District of Columbian, will take offence to my remarks. My only concern is with raising the level of awareness of the working class.

The billionaires are not concerned with the virus, gun violence, unemployment, homelessness, hunger, drug overdoses, or the fact that the infrastructure of the country is falling apart. Their only concern is with one of their own! Donald Trump!

The billionaires consider Trump and his followers to be a threat to their authority, as indeed they are! They are concerned with maintaining their rule! The death and destruction caused by the Virus, as well as the fact that the infrastructure of the country is falling apart, is of no concern to them.

As for those who took part in the largely peaceful protests of January 6, in Washington, hundreds have been arrested and charged with very serious crimes.

The billionaires are also concerned with the fact that the infrastructure of the country is falling apart. More accurately, they are concerned that any maintenance and repairs to that infrastructure, could lead to an increase in the standard of living of the working class!

But as was pointed out previously, according to an article in Business Insider, the capitalists are afraid that such repairs would strengthen the working class, and "enable workers to demand higher wages and force the wealthy to pay a fairer share of taxes".

Those repairs are *not* about to take place! The billionaires are prepared to allow the infrastructure of the country to fall apart! As far as they are concerned, America can sink to the level of underdeveloped countries, of the so called "third world"! They are determined to *not* pay taxes!

The billionaires are about to find out that there are limits! Working people are prepared to tolerate only so much! Those limits are rapidly being approached!

The latest episode of the American judicial system, which is currently referred to as the "gong show" of the court system, is the release from prison of a man who was a convicted sex offender. He was released on a *technicality*, and *not* because he was innocent. Yet he is also a "celebrity", a well-known comedian, and of course quite wealthy. For that reason, he was able to hire high priced lawyers, and they were able to secure his release, upon appeal.

No doubt, this has added to the bitterness and frustration of countless American women, especially those who claim that this same "skinner" (as that is the popular name of sex offenders) has also raped them. Yet the "statute of limitations" protects such criminals, so that they cannot be charged. In fact, possibly fifty women have come forward with this accusation, and very likely there are a great many others. Yet all are being denied justice!

It is also a fact that many women, who have been raped, dare not come forward, if only due to the social stigma involved with being raped. Sad but true!

So many desperate people, so frustrated! All government agencies serve the same class, the billionaires. The only people who care about working people, are other working people.

We can once again refer to a similar situation, which existed in Russia, in 1917. Immediately after the February revolution, in which the Czar was overthrown, the capitalists and landlords were able to seize power, and set up a democratic republic. This was referred to as the Kerensky Provisional Government. This is to say that the common people of Russia, the workers and peasants, had certain democratic rights, if only on paper. In fact, the capitalists and landlords were firmly in charge!

Of course, Lenin was supremely well aware that another insurrection was required. It was necessary to carry the revolution through to its logical conclusion, that of scientific socialism, in the form of the Dictatorship of the Proletariat.

Yet Lenin was also well aware that a great many common people, were not convinced of this! The revolution could *not* be carried through, to scientific socialism, until the vast majority of common people, workers and peasants, *embraced socialism!*

In July of 1917, there was a spontaneous uprising, within Russia. This has gone down in history as the "July Days". This could have led to an insurrection. Yet Lenin was convinced that such an insurrection had little chance of success, as so many common people were *not* convinced of the necessity of a socialist revolution. Lenin called for calm.

Lenin was supremely well aware that it was first necessary to *raise the level of awareness* of the common people, the workers and peasants. They had to be *made aware* that the capitalists and landlords had to be overthrown, the state apparatus smashed, and the capitalists and landlords then *crushed*, under the Dictatorship of the Proletariat.

That was rather difficult, as three quarters of the population of Russia were peasants, and few of them could read. Yet after making a supreme effort, by so many Communist Party members, the situation changed dramatically. This made possible the Great October Socialist Revolution, in October 25, old style calendar, or November 7, new style calendar, of 1917.

Our current situation is similar, although simpler, with fewer classes involved. We too are in the midst of a revolutionary situation. The billionaires have to be overthrown, the existing state apparatus has to be destroyed, and replaced with the Dictatorship of the Proletariat. Scientific Socialism. Yet such an insurrection would almost certainly *fail*, if it took place today, as so many working people are *not convinced* of the necessity of such a revolution!

It was necessary to *prepare* the common people of Russia, for revolution, in 1917. In much the same way, it is now necessary to *raise the level of awareness*, of the American working class. They too, must be *prepared for revolution!*

As documented in a previous article, working people must learn from their own *experience!* They must become politically active! Attempt to "change the system *from within*"! In this way, the workers will learn that the billionaires are in charge, and fully intend to remain in charge!

The more advanced workers must read the fine work of Lenin, State and Revolution, in preparation for the approaching American Revolution. As well, all workers can take part in various Councils (Soviets), which are springing up, all across America. In both cases, workers will learn from *experience!*

Just as Soviets appeared spontaneously, in Russia, at the time of the Revolution, so too they are now appearing in America, under the name of Councils. This has spontaneously given rise to "Autonomous Zones".

The experience of the city of Seattle, was instructive. Capitol Hill declared itself to be an "Autonomous Zone". This Zone was crushed, with considerable brutality. This has taught people that such Zones must not declare themselves to be autonomous. *Not yet!*

The billionaires see these Zones as a threat to their rule! For that reason, they are determined to crush all such Zones.

The working people have responded by establishing Zones, areas of influence within cities and towns, while not kidding themselves that such Zones are autonomous. Instead, they are keeping a "low profile", properly so. The members of such Zones are being armed and equipped, given military training. Preparing for the insurrection, the first step of the revolution.

Now these Zones have to be united, organized into a vast network, a country wide network of working people. Within each Zone, the members have to be trained in the use of various weapons, including clubs, shields, helmets, paint balls, tasers, slings and firearms.

Immediately after the vast majority of the working class is *convinced* of the need to overthrow the billionaires, which will require a revolution, then it will be necessary to *organize an insurrection!* The only people who have proven themselves to be excellent organizers, are the American women. This was proven by the Womens' March on Washington.

In that case, within the space of a few weeks, from the time Trump was elected to the presidency, to the time he was sworn into office, the women organized a nationwide protest. It was possibly the largest, most successful protest movement in the country. Now is the time to build upon that success, to carry the revolution through to its logical conclusion, that of scientific socialism.

Further, the events of January 6, 2021, in the capitol of the nation, referred to as the "one six insurrection", have revealed that the headquarters of the capitalists, the vipers nest of American imperialism, is no where near as well guarded as we would expect. It is clear that a well-organized insurrection could succeed in overthrowing the government, possibly with very little bloodshed.

Ladies, you have done it before, you can do it again! Only this time, no half measures! Go for broke! Seize political power! January 6 was not an "Insurrection", as the capitalists claim! Let them know precisely the meaning of the word insurrection!

In conclusion, may I suggest that the American ladies have their work cut out for them. They have got to learn the ways of the warrior, and that includes the use of firearms. They also have to organize a nationwide network of working people, bringing together the various Zones. All members must be trained in the use of various weapons, including firearms. Preparations must be made for an insurrection, in which the proletariat seizes political power. Allowances must be made for the creation of independent socialist republics.

Young and old should also be encouraged to join the two mainstream political parties, to take part in "changing the system from within". (As if that is about to happen!)

As well, all members should be encouraged to read the most essential works of Marx and Lenin, especially State and Revolution.

Then there is the not so little matter of forming an American Communist Party, Dictatorship of the Proletariat.

Without doubt, that is a tall order. Yet now is the time to prepare for the Dictatorship of the Proletariat. If properly prepared, the revolution, and especially the insurrection, can be quick and painless, almost bloodless. The alternative is anarchy.

www.ingramcontent.com/pod-product-compliance
Lightning Source LLC
Chambersburg PA
CBHW032100020426
42335CB00011B/423